TASTING CIDER

TASTING CIDER

THE CIDERCRAFT® GUIDE
to the Distinctive Flavors of North American Hard Cider

Erin James and CIDERCRAFT® magazine

Storey Publishing

Edited by Carleen Madigan
Art direction and book design by Alethea Morrison
Text production by Erin Dawson
Indexed by Andrea Chesman

Photo credits on page 277

Storey books are available at special discounts when purchased in bulk for premiums and sales promotions as well as for fund-raising or educational use. Special editions or book excerpts can also be created to specification. For details, please call 800-827-8673, or send an email to sales@storey.com.

Storey Publishing
210 MASS MoCA Way
North Adams, MA 01247
storey.com

Printed in the United States by Versa Press
10 9 8 7 6 5 4 3 2

Library of Congress Cataloging-in-Publication Data

Names: James, Erin, 1985– editor. | Cidercraft Magazine.
Title: Tasting cider : the Cidercraft guide to distinctive flavors of North American hard cider / [edited] by Erin James and Cidercraft Magazine.
Description: North Adams, MA : Storey Publishing, [2017] | Includes bibliographical references and index.
Identifiers: LCCN 2016059753 (print) | LCCN 2017019160 (ebook) | ISBN 9781612128382 (ebook) | ISBN 9781612128375 (pbk. : alk. paper)
Subjects: LCSH: Cocktails. | Cider—North America. | Cider industry—North America—Directories. | LCGFT: Cookbooks.
Classification: LCC TX951 (ebook) | LCC TX951 .C56 2017 (print) | DDC 641.3/411—dc23
LC record available at https://lccn.loc.gov/2016059753

This book is dedicated to all those who make the cider world go 'round — from the makers to the readers, the novices to the enthusiasts, and everyone in between. Raise a glass and drink cider!

CONTENTS

INTRODUCTION

Those of us at CIDERCRAFT magazine get called a lot of names, but first and foremost, we are cider drinkers. Enthusiasts, journalists, writers, advocates, promoters, sure — but the consumption of the golden liquid in that glass is the reason why we're all here, why we wake up in the morning, and what we hold at night. For years, we've been covering the trends and growth of this burgeoning beverage, yearning and anticipating for cider's time and place to shine. That time is now, that place is here, and we are CIDERCRAFT — North America's first print publication dedicated to hard cider. And this is our book about cider.

Cider is as American as apple pie. Not the garden-fresh apple juice from a pumpkin patch on a chilly, plaid-clad October afternoon or even the sugary apple brew of the 1990s. But *cider* — the drink that helped build North America. Like our magazine that launched in early 2014, this book gives a snapshot of that forgotten history, of the basics on how it's made, and a general tasting tutorial with recommended glassware, while digging into the variations of the drink from the arcane and the barrel-aged ciders to the avant-garde and the delicate perry. We collaborated with bartenders and chefs across the continent for recipes that mix, cook, and pair with cider, and we circled around the campfire gathering stories from North American producers who share their passion with such enrapturing vigor that you can't help but pour more of their juice in your glass.

Raise it up! This is *Tasting Cider*, a toast to North American hard cider.

CHAPTER 1

THE CORE ESSENTIALS

The United States of America was established upon man's basic rights in the New World: life, liberty, and the pursuit of alcoholic beverages. The influencers of the nation that became sovereign in 1776 were also likely looking forward to what else the fall harvest would bring — cider. Thomas Jefferson planted nearly 50 trees of his beloved Albemarle (Newtown) Pippin cider variety on his Virginia estate. George Washington's 1758 election campaign to the legislative assembly poured roughly 43 gallons of hard cider and beer to prospective voters. John Adams was an ardent apple-a-day consumer, and he took his servings in the form of cider. And some of Benjamin Franklin's best-known one-liners address the fermented fruit juice: "It's indeed bad to eat apples; it's better to turn them all into cider."

Early Canadian settlers witnessed the same cultural insurgence, with saplings and seeds coming mainly from France to "New France" (modern-day Québec). The first recorded cidery was established in the mid- to late seventeenth century on an orchard near present-day Montréal. Largely produced by those of faith, cider in Canada was a mainstay in the New France Catholic Church, even when the temperance movement came knocking in the early twentieth century — a solid 78 percent of Québécois voted for "beer, ciders, and light wines" to be excluded from the list of prohibited alcoholic beverages. This, unfortunately, did not stick, and cider remained illegal to sell until 1970 due to an oversight in the Alcoholic Beverages Act that left the drink out of the original legislation that ended Canada's prohibition.

Courtesy of the temperance movement, the prohibition of alcohol, and various agricultural changes, the once-booming cider industry fell to the wayside across North America. Craft producers were bottling small commercial allotments in the 1970s and 1980s, but it wasn't until 2011 that cider sales started taking off again, making cider the fastest-growing alcoholic beverage in North American history. Perry — the fermented drink made from pears that slips into the category and statistics of cider — is also seeing its own revival.

By and large, cider hadn't seen this sort of success since it was first sipped by the founding fathers.

The Jodoin family, of Québec's Cidrerie Michel Jodoin, harvests from the estate orchard in September 1950.

Today, cider is making an astronomical comeback, and consumption of this beverage continues to rise at a staggering rate. That's something ol' George Washington would be proud of. From sea to shining sea, cider holsters its history and spouts sensational stats; here are a few fun facts to juice up your cider knowledge from the start.

- **Cider isn't gender specific.** Men and women share a piece of the pie, with half of the drinkers being male and the other half female, unlike any other beverage.

- **Cider isn't noooooarily sweet.** Apples might be high in sugar content, but the process of fermentation turns those sugars into alcohol, resulting in a high-octane sipper comparable in alcohol by volume (ABV) to the popular and potent India pale ale style of beer, averaging 6.5 percent and higher. After fermentation, the cider is typically dry. From that point, producers are able to add juice, sugar, or other sweeteners to sugar it up.

- **Cider drinkers are often converted beer geeks.** Drinkers of craft and imported beer are among some of the first to test the unknown and are eager to continue testing, so cider welcomes many of its newcomers from the beer world.

- **Québec is the most densely populated cider area on the continent.** With 70+ cideries in the province and growing, the majority of the producers make at least one ice cider. The orchard-rich and scenic Montérégie region accounts for more than 60 producers.

- **There are at least 10,000 apple orchards in North America.** The United States hosts roughly 7,500, while approximately 2,500 are cultivated in Canada.

- **The two oldest continuously operating cider mills in the United States are less than an hour away from each other.** B.F. Clyde's Cider Mill in Old Mystic, Connecticut, has been running the oldest (and now only) steam-powered cider mill in the country since 1881, and the Old Cider Mill has been functioning since the 1870s in nearby Glastonbury.

- **Cider is made in nearly every American state and in four Canadian provinces.** Fruit and juice might be sourced elsewhere, but these areas are bottling and packaging their own product in the name of cider.

A SNAPSHOT OF CIDER HISTORY

If history shows anything, it's that mankind is fond of its drink. Whether man was fighting for the right to party against a wave of oppression, temperance, religion, agricultural cataclysm, and so on, drink still found a way back into the hands of humans. Dating back to the Stone Age, fermentables have been a means to sustain against hunger and even from disease and insanity. Basic fermentation science shows that if there is the will of yeast and offering of sugar (like those found in fruit), there will be a way for alcohol, and that condition has not changed with time.

ORIGINS OF THE APPLE

Just as the apple doesn't fall far from the tree, the origin story of cider begins with the apple itself. The exact provenance of the apple is vague, although most historians believe the unearthing of the modern apple's roots showed up millions of years ago in the Dzungarian Ala Tau, the mountain range separating Kazakhstan, Kyrgyzstan, and China.

The fruit, following the pilgrimage of civilization, moved farther eastward over time and westward by 6500 B.C.E. from Egypt and Israel to Greece and Spain. An ancient tablet found in Mesopotamia dating to around 1500 B.C.E. served as a receipt from an apple orchard sales transaction and literally puts apple orcharding in stone. Sometime in the thirteenth century B.C.E., pharaoh Ramses of Egypt ordered apples to be grown along the Nile for aesthetic beauty and to please his palate. In 323 B.C.E., Theophrastus, Greek botanist and successor to Aristotle, scribed of six different apple varieties, depicting the growth and grafting of apples. Roman philosopher-politician Marcus Cicero pushed for the cultivation of apples throughout the empire and, although not directly related but relevant to his love of drink, was quoted as saying, "Almost no one dances sober, unless he is insane."

Three nymphs in the Garden of the Hesperides guard the sacred golden apples for the Greek goddess Hera.

Through the practice of agriculture and the process of cultivation, the apple's popularity increased and took on roles outside of fruit, giving it a personality and purpose of its own via folklore and religion. In Christianity, the apple is cast as an infamous character in the origin of sin with Adam and Eve. (Interestingly enough, modern research has revealed that it might not have been the apple but a fig that caused the fall of mankind.) Greek mythology tells of the goddess Hera's Garden of the Hesperides, which boasted of an apple tree producing golden fruit that offered immortality. Norse legend has it that the gods also achieved eternal life by eating special apples from the orchards of the goddess Iduna. Celtic folklore includes the story of a man who receives an apple that feeds him for a year. Seemingly ageless and spanning ideologies, the apple embodies love, sexuality, and everlasting life.

TRAILBLAZING EUROPEAN CIDERMAKERS

Cultivation of apples was widespread by the time the Romans invaded Britain in 55 B.C.E. and tasted their first cider. Some accounts testify that even emperor Julius Caesar himself was a cider enthusiast. The names for the fermented apple juice were tossed about by different cultures, in different tongues. A common belief among cider drinkers and researchers is that the Latin word *sicera*, along with the Hebrew word

shekar, is the driving dialect that landed on "cider," although both words were used to describe any alcoholic drink that was not wine. In 60 B.C.E., the heavily bearded drinks enthusiast and Greek philosopher Strabo made mention of the fermented drink *zythos* in Asturias of northern Spain, which Spanish cider proponents believe was in reference to cider. Typically acknowledged as the world's first cidermaking regions, Spain's Asturias and Basque cider principalities have stood the test of time and are two of the strongest cider cultures to date.

Roman philosopher and author Pliny the Elder (who receives modern acclaim mostly for his chronicling of booze over his prolific credo on life) writes in his first century C.E. book *The Natural History* that "wine is made, too, of the pods of the Syrian carob, of pears, and of all kinds of apples." Known for enjoying a tipple or two, Pliny speaks of fermentation to apple wine and to cider, although it had not received recognition as its own moniker.

Historians believe the Frankish emperor Charlemagne might have been the first to give cider a specific name — *pomatium,* and *pyratium* for perry — outside of the miscellaneous beverage umbrella titles. In the sixth and seventh centuries, Charlemagne's empire ranged from northern Spain to France and Germany, countries that were then as known for cider as they are today, including Normandy in France and Asturias

Historians believe the Frankish emperor Charlemagne might have been the first to give cider a specific name — *pomatium*, and *pyratium* for perry.

in Spain. Documentation does not credit which culture pushed cider the furthest, but Charlemagne's empire did make a sizable footprint in the history and culture of the drink.

Thanks to the Norman Conquest (the eleventh-century invasion and occupation of England by a joint effort army of Norman, Breton, and French soldiers), cider and perry were flourishing as potent potables in Western Europe. Changes in weather patterns had brought a crippling demise to grape growing in both France and England, and the Normans quickly helped replenish the area's liquor cabinets with cider by planting hardy cider apple trees from France and into England. Cuttings of popular varieties were passed along, thus spreading cider throughout Western Europe.

HOW TO SAY "CIDER" IN

SPAIN: SIDRA. One of the origins of cider, including both the Basque and Asturias regions, *sidra* is poured from a bottle held high above the head of the *escanciador*, a cider expert skilled in the art of pouring, and into the glass held below the waistline. The dry, still cider is aerated as it soars from the bottle to the glass and creates natural carbonation.

FRANCE: CIDRE. The ancient beverage hails from the equally old areas of Normany, Brittany, and the French Basque country. Styles range, but oftentimes *cidre* is seen with more ripe fruit character, less alcohol, and a sweeter, robust palate. Sparkling *cidre* goes by the name of *cidre bouché*, and levels of sweet range from *doux* (sweet) to *brut* or *sec* (dry).

UNITED KINGDOM: CYDER. The people of the United Kingdom drink more cider than the rest of the globe, with the sprawling orchard coverage to prove it. They also rank as the longest, continuously cider-drinking culture on record: since the first century B.C.E. Roman expansion. Outside of cyder, the tipple also goes by the name of "scrumpy" when it's produced in a rustic, farmhouse style.

GERMANY: APFELWEIN. Easy to translate for English speakers, *Apfelwein* is also known as *Ebbelwoi*, *Äppler*, *Stöffsche*, and more. The cider is made largely from culinary or dessert apples, and the drink itself is growing in demand, with more than 60 producers in the country.

AUSTRIA: MOST. Lower Austria's Mostviertel cider region is home to Europe's largest continuous area of pear orchards, and the 200-kilometer themed Cider Route showcases the famous region's *Most* and *Mostheuriger*, the farm-based inns for guests.

Using a traditional pouring technique for Asturian cider in northern Spain, this server holds the bottle high above the glass below to elicit natural carbonation in the still drink.

THE EARLY EUROPEAN CIDER MOVEMENT

By the turn of the twelfth century, cider production extended from Spain and France to northern England. Orcharding was a way of life for these countries and cultures, nourishing stomachs with apples in the diet and spiking pub tankards with the sweet swig of alcohol. The production of the fermented beverage was also a means of preserving the fruit — the apple in liquid form could hold longer in a barrel or vat than intact apples in a cellar. More often than not, the naturally occurring yeasts within the apples (both on the skins and in the flesh) would turn the juice into cider within days of containment — by lucky, heady happenstance.

Because many of these original European cider orchards were grown from apple seeds and not rootstock grafts, their fruit was an amalgamation of varieties, resulting in new apple types. To reproduce a particular apple cultivar, growers had to (and still do) graft it onto the rootstock of another tree, as a tree started from an apple seed does not produce the same fruit as the parent plant and will result in a novel variety. This practice of cloning ensured the regular planting of popular apples, and by the late 1500s, France had slapped designated titles on more than 60 apple varieties.

In England, far and few between was a farm that didn't grow, crush, and ferment its own apples for cider. By the eighteenth century, the beverage was a conventional form of payment for farmworkers, who averaged three to four pints for daily wages. In some cases, laborers were even evaluated on how well they could hold their alcohol.

The English and French held their cider so well, in fact, that they brought it to the New World. And that's where we come in.

Before electricity, there was horsepower. In France and elsewhere in Europe, apples were crushed for cider using a horse-drawn mill.

FROM PURITANS TO
JOHNNY APPLESEED

In tow with the first English and French settlers, cider found itself at the center of the countries' foundations, long before the United States' proclamation of independence or the creation of the Canadian Confederation. By some accounts, it was a large iron screw from a cider press on the *Mayflower* ship that helped hold up a supporting beam after a storm badly damaged the crucial plank.

Once in the New World, it took the Puritans all of nine days before they were planting the apple seeds they carried across the pond with them in 1607. William Blaxton (also written Blackstone) takes credit for planting the United States' first cultivated apple variety as well, the Rhode Island Greening (previously called Yellow Sweeting) shortly after his arrival in 1623. Rightly so, commercial cider production started soon after.

The first apple tree planted in Canada was in 1617 Québec, by French apothecary Louis Hébert. As one of the first settlers to the area, he called the shots on what was planted where, putting this inaugural tree on the site where Québec City had been founded 11 years earlier. A majority of the early Canadian frontiersmen and women were Normans, a culture that embraced cider drinking as akin to eating breakfast. Records for Canada's first cider production show up later in the same century in modern-day Montréal, Québec.

A THRIVING CROP FOR THIRSTY SETTLERS

Along North America's eastern coastline, many of the agricultural cuttings and seeds that traveled across the Atlantic failed to root and prosper. Farmers who saw crops collapse began to rethink why they made the move in the first place. But the apple was a survivor. It would replicate and reinvent itself as something new with each seed, forming unique varieties bred for the colonies' specific climates and fostered to thrive.

Rhode Island Greening is one of the oldest American apple varieties, cultivated since the 1600s.

The first documented blooming orchard was in Massachusetts Bay Colony, just outside present-day Boston, where the oldest plantings of apple varieties, like Roxbury Russet and Rhode Island Greening, hail from. One of the earliest nurseries on record called the same colony home, with its apple trees allegedly planted for alcohol production and sold to colonial arrivals to take to their new homes. By the mid-1600s, the newly established authorities were drawing up laws and regulations on booze, and hefty fined citations were dropped onto those colonizers who sipped their cider with a heavy hand — the original tickets for public drunkenness.

The apple served its purpose not only as an edible option for aspiring farmers and thirsty home brewers, but once in fermented form, it was an alternative to oftentimes suspect drinking water in the new colonies. Offering sustenance as well as a source of fruit, hard cider was a family tradition. Parents and children, young and old, would drink cider for its nutritional (and intoxicating) benefits. Reports show that an average family of six would drink nearly 90 gallons of cider per year, amounting to about 15 gallons per person.

Cider's success in the New World was thanks to its traditional role in the motherlands of England, France, and Spain but also because of the efforts of the founding fathers of the United States. From George Washington's 1758 cider campaign and Thomas Jefferson's sweeping plantation home that featured a personal orchard of Albemarle (Newtown) Pippins, cider was a political move forward.

John Chapman (more widely known as "Johnny Appleseed") demonstrates how to plant an apple tree.

SOWING SEEDS FOR CIDER

And then there's John "Appleseed" Chapman, an accidental botanist who scattered apple pips from cider mills in Pennsylvania across the emerging frontier, west to Illinois, south to West Virginia, and north to Ontario. He helped establish nurseries and orchards in the nineteenth century. Contrary to popular belief, Chapman wasn't spreading Red Delicious seeds barefoot and whistling a Disney-honed tune — he was laying out trees for cider apples and spreading this historic beverage movement.

Until the temperance movement gained steam in the nineteenth century, North American apples were grown for cider production, not culinary use. John Chapman died in 1845, and many of his orchards and apple varieties planted for cider eventually left with him. Trees producing inedible (sour, bitter, tannic) cider apples were chopped down by government agents, even though Prohibition did not halt the growth of corn (whiskey), wheat (beer), and grapes (wine). Cider itself was quickly forgotten. Until now.

The rise of cider today in North America isn't because the drinking water is polluted or the beverage market needed more gluten-free options to quench the thirst of conscientious CrossFit fanatics. Cider has stepped into the limelight because the consumer demanded it. An educated, eager craft drinker has asked for something beyond the hard lemonades and spiked root beer. The craft drinker requests a drink as complex as she is, and cider combines fruit, history, and flavor in one intricate, storied tipple.

HIP ON HORTICULTURE

The lexicon for cider is enough to swallow on its own before adding in horticulture terms, varieties, and more. Key words are always good to have ready when speaking with the natives, and below are the basics on what and how to propagate an apple tree.

ROOTSTOCK. Also known simply as "stock," this is a root system, or the bottom/primary portion of the tree, onto which other varieties are grafted.

SCION. A dormant, detached shoot or twig from one plant that is attached to the rootstock of another plant. The term "scion wood" is also used.

GRAFTING. The technique used to join the rootstock of one variety to the scion of another. Since trees grown from apple seeds do not typically produce the same fruit as the parent plant, grafting allows for the reproduction of choice apple varieties.

SEEDLING. A young tree grown from seed. Apple trees grown from seed almost never yield fruit with the same attributes as the parent tree. This random and spontaneous production of new varieties has led to a wide selection of apples throughout Europe and North America.

APPLES: FROM FRUIT TO JUICE

Speaking in sheer numbers, much of the cider made in North America is fermented from culinary or dessert eating apples, with the bulk of apple tonnage coming from Washington State, New York, and Michigan. More than 7,500 apple growers span the United States, cultivating nearly 200 varieties on approximately 328,000 acres. Across Canada, apples are planted on about 50,000 acres through 2,500 growers found mostly in Ontario, Québec, and British Columbia. The United States is the world's second-largest producer of apples, just behind China.

Somewhat influenced by the explosion and market ultimatum for bigger, better, and more Red Delicious apples to supply grocery store demand, the dessert apple trade still reigns supreme on these orchards. The lion's share of trees from the earlier colonizers are long gone, but a few historical varieties have made their way back into the mainstream. And an increasing number of orchardists are planting imported saplings of traditional French and English cider apples, along with heirloom cider apples from New England, New York, Virginia, and even the West Coast.

NORTH AMERICAN HEIRLOOM APPLES

While a small number of orchards today are planted with the classic varieties considered ideal — by traditional European criteria — for cidermaking, producers across North America are discovering tenured and heritage varieties that are long established to American cidermaking. Esopus Spitzenburg, Golden Russet, Northern Spy, Gravenstein, Baldwin, Rhode Island Greening, Albemarle (Newtown) Pippin, and Arkansas Black are among many now-domestic apple options that offer similar characteristics to "the classics" with a truly North American touch.

Distinctively aromatic and flavorsome while driving acid forward and fashioning shape and body from simple tannic structure, these "heirloom," "heritage," or "antique" apples are seen mostly on the East Coast and the Great Lakes region but are popping up in historical homestead properties and estates of cider-focused orchardists across the United States and Canada.

"Give me yesterday's bread, this day's flesh and last year's cyder."

— Benjamin Franklin

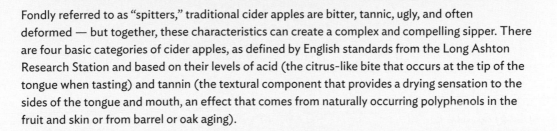

THE CLASSES OF CIDER APPLES

Fondly referred to as "spitters," traditional cider apples are bitter, tannic, ugly, and often deformed — but together, these characteristics can create a complex and compelling sipper. There are four basic categories of cider apples, as defined by English standards from the Long Ashton Research Station and based on their levels of acid (the citrus-like bite that occurs at the tip of the tongue when tasting) and tannin (the textural component that provides a drying sensation to the sides of the tongue and mouth, an effect that comes from naturally occurring polyphenols in the fruit and skin or from barrel or oak aging).

SWEETS. Low in acid and tannin, these apples are as advertised: sweet and even-keeled. Traditionally, these apples are used to blend into ciders with stronger varieties to balance out otherwise extreme acidity and tannin, although more modern cideries are using sweets for a juicy base. Common varieties include Gala, Fuji, and Honeycrisp.

BITTERSWEETS. A classically used category for English ciders, these apples are low in acid and high in tannin (hence the name). The tannins offer astringency and dryness, with varying levels of both depending on the variety. Common varieties include Yarlington Mill, Dabinett, and Tremlett's Bitter.

SHARPS. These apples are ample in acidity — sharp and fierce prima donnas — and are typically more difficult to find, as a little bit goes a long way. Common varieties include Granny Smith, Rhode Island Greening, and Winesap.

BITTERSHARPS. High in acid and high in tannin, these apples exhibit traits similar to black tea or lemons and are often seen in blends with bittersweets. They don't mess around and are revered for their intensity. Common varieties include Kingston Black, Stoke Red, and Porter's Perfection.

Fuji

Winesap

| SWEETS | BITTERSWEETS | SHARPS | BITTERSHARPS |

On all sides [Ichabod] beheld vast store of apples; some hanging in oppressive opulence on the trees; some gathered into baskets and barrels for the market; others heaped up in rich piles for the cider-press.

— Washington Irving, "The Legend of Sleepy Hollow"

PRESSING CIDER

Whether it's made from dessert, heirloom, or European cider apples, hard cider is produced by crushing the fruit and fermenting the juice. The cider presses or mills that are used to crush the fruit vary in size, stature, and style. The most historic option is the "rack-and-cloth," a seventeenth-century method that stacks ground chunks of apples in racks that are closed by cloths and pressed to expel the juice. The most modern presses come from the wine world and

are built with hydraulics and stainless steel for large-scale production.

Once the ripe, sugar-laden, sun-soaked fruit is ground or crushed to release its juice, yeasts (whether added or naturally occurring on the apple skins) transform the sugar from the juice into alcohol. This is fermentation: the biological process in which sugar is converted by yeast into carbon dioxide gas (CO_2) and ethyl alcohol (in this case, cider). Science!

Apples for cider must first be crushed and pressed to expel the juice. Although the "rack and cloth" method shown is centuries old, the hydraulic press that powers it is a modern addition.

HOW TO TASTE CIDER

Describing cider beyond its inherent apple aromas and flavors can be cumbersome and not always necessary. You like it? Great, drink it. Beyond finding a drinking vessel nearby and pouring it in, cider falls in line with wine and beer as far as understanding and appreciating the beverage — something that comes with the celebrated practice of drinking.

CIDER-CENTRIC GLASSWARE

Like its beverage brethren, cider has its "ideal" glassware, a format that helps express the hidden and subtle aromas. In cidery tasting rooms across the continent, cider is served in a variety of glasses, but most wineglasses, either stemless or stemmed, will fit the bill as the appropriate choice. A wineglass with a larger bowl (everything but the stem) allows for swirling and coaxing out aromas, while

a narrowly tapered beer glass will aid in angling the aromas toward your nose. The thinner the material of the glass the better — then you taste the cider, not the material of the drinking vessel.

The manufacturing industry is catching on. More and more cider-centric glassware is entering the market, like Libbey's Hard Cider Glass (this page, below left) and 33 Books Co.'s Original Cider Tasting Mug (below middle), a pub-inspired, wide-open ceramic tankard for optimal cider sipping. With national stores like Williams-Sonoma, Crate and Barrel, and Macy's stocking their shelves with brands specific to cider for savvy consumption, the options are growing on trees.

At the first shrill notes of
    the pipe
I heard a sound as of
    scraping tripe
And putting apples,
    wondrous ripe
Into a cider-press's gripe.

— Robert Browning, "The Pied Piper
    of Hamelin"

TEMPERATURE, TASTE, AND SMELL

Before the cider makes it to the glass, be sure to note the serving temperature; this can help or hinder the tasting experience. Dry and tannic ciders can handle room temperature, while sweeter ciders should be served colder. If the cider is at a loss for aromatics, it could be too cold. If it shows too alcoholic, chances are you are tasting it too warm. At the end of the day, though, taste is subjective, and the cider should be consumed within the specifics of your personal preference.

Olfactory research has shown that 80 to 90 percent of that taste comes from smell, so don't be shy about shoving your nose into the glass to take it all in. Like wine or beer tasting, swirling the cider is recommended to release these aromas. Before taking a sip, try swirling and sniffing a few times, to get the full effect.

If you're looking for a complete tasting experience, consulting a flavor wheel will help you identify what is in the glass (see page 20). Start from the inside and move out — if you're finding sweet aromas, are they honeyed, syrupy, vanilla-y? If it's spicy, is it nutty or grassy, woody or resinous? Splash a glass, spin the wheel, and become bilingual in the aromatic and flavor profiles of cider.

THE SWEETNESS SCALE

In the most basic explanation, ciders are categorized into four levels of sweetness. Although during fermentation most cider is fermented to complete dryness (in which the yeasts have consumed nearly all the sugars in the cider), sugar, unfermented juice, or any other natural sweetener can be added back in to increase the sugar level. The exception is ice cider, which is made from two different methods of freezing the fruit or juice, but which also falls into the sweet category.

DRY

Both in still and sparkling renditions, these ciders tend to have higher tannin and bolder acid profiles that reveal the stark and mineral side of the apple. The residual (remaining) sugar in the cider is typically less than 0.5 percent gram per liter. A prime example is the bone-dry Winesap/Albemarle (Newtown) Pippin blend called Terrestrial from Castle Hill Cider in Keswick, Virginia.

Olfactory research has shown that 80 to 90 percent of that taste comes from smell, so don't be shy about shoving your nose into the glass.

OFF-DRY	SEMISWEET OR SEMI-DRY	SWEET
Between 1 and 2 percent residual sugar, these ciders tend to exhibit a fuller body than bone-dry ciders, with more fruit-forward aromatics and flavors while maintaining some tannin and ample acid. The Legend, a combination of heritage and dessert varieties from Rootstock Ciderworks in Williamson, New York, represents nicely for this level.	Somewhat of a middle ground between dry and sweet — between 2 and 4 percent residual sugar — these full-bodied, robust ciders are on the softer side of tannin and acid while showcasing juicy fruit and easy drinkability. Most mass-marketed, large-scale commercial ciders fall into this category. The Gumption, a mix of cider and dessert apples, from Woodchuck Hard Cider in Middlebury, Vermont, falls in this range.	Once the cider hits the 4 percent residual sugar mark, it is often classified as sweet. With ice ciders also falling into this category, sweet or dessert ciders often average anywhere between 10 and 20 percent residual sugar and are reminiscent of a dessert wine. A choice option is Frost, a dessert cider from Tieton Cider Works in Tieton, Washington, that is a late-harvest blend of Jonagold, Pinova, and Winter Banana apples.

HOW TO TASTE CIDER

19

THE FLAVOR WHEEL

Courtesy of cider expert Andrew Lea, and based on testing done at the Long Ashton Research Station in England, the information in this flavor wheel classifies cider into familiar senses — aromas, tastes and textures — to aid in interpreting the cider in your glass. Use it as a guide to deciphering the aromatic and flavor profiles of your cider, remembering that the route to tasting cider is subjective and you're behind the steering wheel. Take hold and drive in the direction that fits your preferences.

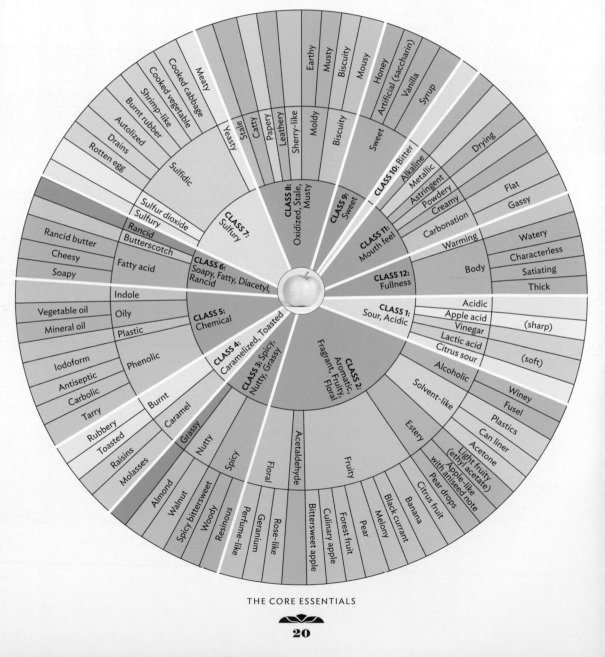

CIDER 101: A GLOSSARY OF TERMS

Talk like a pomme enthusiast — learn the basics of cider and perry terminology to feel confident walking up to a bar or retail shelf, picking the right prize, and reaping the boozy benefits of your selection.

ABV. Alcohol by volume, a guiding measurement used throughout alcoholic beverage production to calculate how much alcohol is contained in a certain amount of volume of alcohol. Denoted as a volume percentage.

Acidity. The sharpness or briskness found on the top of the tongue when drinking cider, a balancing flavor component based on the amount of malic acids in the beverage.

Adjuncts. Added matter to a fermenting or fermented cider, like hops, spices, or botanicals.

Astringency. The dry, mouth-puckering effect from acid or tannins; a positive influence when balanced in a beverage.

Barrel aging. Either during or after fermentation, a cider can spend time in oak barrels to mature and take on the aromas of the barrel.

Bottle conditioned. A secondary fermentation process that happens in the bottle when yeasts are left in the cider after bottling to mature the cider and induce natural carbonation.

Brix. The sugar content of either an apple or the unfermented juice.

Carbonation. The resulting bubbles in cider from bottle conditioning or injecting carbon dioxide gas into the fermented juice.

Barrel aging mellows the cider and infuses it with the aromas of the cask.

Ciderkin. A lower-alcohol beverage made from rehydrating, pressing, and fermenting the leftover pomace (apple skins, flesh, stems, and so on).

Concentrate. The concentrated form of apple juice that is stored and rehydrated in order to ferment.

Draft. Cider that is held in a keg, instead of a bottle, and is served through a tap.

Fermentation. When yeast converts the sugars in juice to alcohol.

Finish. The final aftertaste left in the mouth after swallowing a beverage.

Heirloom, heritage, or antique apples. Older apple varieties, either native to or longstanding in North America. Three species of crabapple (and their cultivated varieties) are the only apple trees native to North America.

Inoculate. The process of adding active yeast cultures into unfermented juice to start the fermentation.

Keeving. A traditional French process that removes nutrients from the juice in order to slow and elongate fermentation, sometimes finishing without fermenting to dryness.

Cidermakers mill the apples (on the right) before pressing them (on the left).

Malic acid. A principal acid in ripened fruits, providing a sharp, astringent, or "green" taste. This often-negative attribute is tamed through malolactic fermentation, where the tart malic acid is converted to smoother, softer lactic acid.

Malus. A genus of deciduous trees, including the species of *Malus domestica* (a.k.a. *Malus pumila*), the orchard apple tree.

Milling. The grinding or crushing of apples to prepare them for the press.

Mouthfeel. The physical perception of how the cider is sensed on the palate and in the mouth, such as texture and weight.

Palate. Literally the roof of the mouth; figuratively (and in terms of cider), an appreciation for taste. Often used to describe flavor profiles.

Pasteurization. A simple heating process (to at least 160°F/70°C) that destroys pathogens and ensures the safety of the beverage.

pH. A numeric scale used to quantify the specific acid of a liquid.

Pomace. The solid remains of wine grapes or apples after pressing; typically composed of stems, seeds, pulp, and skins.

The aftermath: pomace piles up after pressing.

Pressing. The production step in which milled fruit is squeezed to release its juice.

Scrumpy. A term for rustic ciders that come from England's West Country; scrumpies are typically produced in small batches, using traditional methods.

Sparkling. The same as carbonation, a bubbly sensation to the cider and on the palate. Either achieved by forced (injected) carbonation or by secondary fermentation in the bottle.

Still. A cider without any carbonation, natural or forced.

Sweet cider. Nonalcoholic cider, also known as apple juice.

Tannin. An astringent, natural phenolic compound that is sensed by dryness on the sides of the tongue and mouth, either as a result of the apple used or aging in oak.

Terroir. From the French word *terre* meaning "land," this term is used to describe a specific area's identity and sense of place, and how it affects the agriculture and the resulting flavor of the food and drink produced there.

Yeast. A microscopic fungus that converts sugar into alcohol and carbon dioxide. Some cidermakers use wild yeasts from their natural surroundings, while others add a commercial or industrial strain (like an ale or Champagne yeast). To each their own; different strains produce different results. A white wine yeast, for example, might produce a lighter, more delicate cider, and a Belgian ale yeast might produce a bolder, spicier cider.

The ciders at E.Z. Orchards represent the terroir of their orchard in Salem, Oregon.

FROM
BRANCH
TO
BOTTLE

The most difficult aspect of cider isn't the drinking or even the making, but the defining of the beverage. It seems almost contradictory that North America's oldest drink is still figuring out how to interpret itself. However, it is the diversity of the language of cider — the range of style variations and forms the fermented apple can take — that make it a unique libation, unlike any other.

Sure, you could distill a shoe and create a spirit. Beer has reached past its barley and wheat confinements and now comes in gluten-free manifestations. But cider must come from the apple; be it fresh juice or concentrate, the derivation is fundamentally the same.

With the basics laid out, cider is still in its modern resurgence. The cider prior to Prohibition is different from the cider available today, from the apples to the production methods, and the terms used to define the beverage category are not standardized. In speaking with producers throughout North America, we came to a conclusion: Who are we to classify ciders when those making the drink don't?

FREESTYLE CIDERMAKERS

In North America, land of innovation and avant-garde, there is a melting pot of traditions that frees makers to take full rein on their creations. Without centuries of continuous production, regulation, and establishment like cider's European makers, North American cidermakers are not bound to one method or another.

Even history has had a hard time nailing down specifics, with many ciders being made from a single orchard, oftentimes without any more information than the batch being a field blend of whatever was grown on-site. Regional variations of cider have arisen, but even though the English are known for dry, tannic, and still cider; the French for sweeter, riper styles that are naturally sparkling; and the Spanish for tart and acidic, those terms do not necessarily define the "style" itself.

A founding board member of the United States Association of Cider Makers and co-creator of the annual national Cider Conference, James Kohn shared with us his four different "labels" for cider and perry, based on basic production facts: geography, types of apples or pears, the yeasts used, types of fruits added (if any), and adjuncts added such as hops or spices (if any). Under these labels, Kohn identifies the ciders of his Wandering Aengus Ciderworks as from Oregon, using heirloom apples, fermenting with wine yeasts, and without fruit or adjuncts. He believes defining a label or style could be as simple as that.

"It's mostly production-based labels, and those things aren't controversial," he says of his classification, noting how wine is categorized in a similar sense on the grocery shelf. "This is a guide to consumers and retailers; it gives an identity of what you're drinking and why it tastes that way."

PURIST OR PROGRESSIVE?

In the United States and Canada, there seems to be two schools of thought on cider and cider production. One, the more traditional and purist circle, sees cider on the straight and narrow: from the orchard, from specific apples, and made a certain way. The other, the more progressive and experimental mind-set, takes what Mother Earth and the market gave them and ferments that, be it dessert apples or apple concentrate, and distinguishes it with other components (hops, spices, other fruit). Producers can alternate between the two, depending on what they are making, but they more or less stick to their respective conclusions for their individual production. Like opposing political parties during election season, both have valid points, strong stances, and vigorous momentum — but neither one is necessarily the right or wrong candidate to get the job done. Both promise to be the best they can be and, at the end of the day when all you need is a drink, that's worth a vote.

Without centuries of continuous production, regulation, and establishment like cider's European makers, North American cidermakers are not bound to one method or another.

CIDER
VARIATIONS

Through the rigorous research and arduous practice known as "tasting" (or drinking), we have settled on 11 variations of cider for the purpose of this book: orchard based, modern, single varietal, dessert, hopped, rosé, fruit infused, barrel aged, spiced and botanical, and specialty, with perry to boot. These variations are more commonly seen in the North American marketplace and are ones you can buy now. Avoiding the hot-button term "style," and acknowledging that a cidery is not limited to a specific variation, the following ciders are a starter set for the consumer — the basic genres of the great North American drink. Far from definitive but thoroughly explored, North American cider flexes its muscles and shows its brawny span.

ORCHARD-BASED CIDERS

It's a rarity to find an American cidery that doesn't preach about the "tradition" of cider in some verbiage or another. Forefathers, presidents, generational elders, and estates are all name-dropped to showcase heritage, legacy to Thomas Jefferson's favorite variety (he had several), or one of New England's oldest apples. Traditional production can be reminiscent of the Colonial era or of Western European techniques. Or, like today, simply American or Canadian, based in the orchard and all about the apple.

THE HISTORY

Definitions for what "tradition" or "traditional" is in the sense of North American cider are sparse and often backed with colorful language. Many producing ciders that could fall into this category are averse to the label, or any label even, finding sticker solace in the term "orchard based" over anything else because — when the apples are pressed and the juice is fermented — that is where the cider came from.

The chronological arithmetic is simple. First came the apple, then came the cider. But before the apple, there was a tree, and that tree was in an orchard. Orchard-based production can mean many things, but most producers find the tag being used for ciders that have minimal travel time spent between tree, press, and fermenter. With apples at the core of the production — the variety, the blend, how it's fermented and aged — the purity of the fruit is priority number one.

THE BREAKDOWN

The cidermakers and orchardists behind this practice do not always respond to designations like "traditionalist" or "purist," because that isn't always the case. Some orchard-based producers can ferment a mean single-varietal Kingston Black bittersharp cider, and they can also blend in whole blackberries and Amarillo hops to some Baldwin juice and capture the full field of the orchard's agricultural digest. The apples, for the most part, are varieties of "intention": you won't often see Fuji as the base of these ciders, but you are likely to spot traditional European cider varieties like Ashmead's Kernel and Chisel Jersey or American heritage fruit like Golden Russet and Hewes Crab.

"I would categorize my ciders as dry and orchard based," says Stephen Wood, cidermaker and orchardist of Poverty Lane Orchards and Farnum Hill Ciders. The cidery sports the words *traditional cider* on the label, but de facto of Wood's original intention behind the term. "The 'tradition' we were appealing to was the idea of Americans drinking cider *at all*. We had no cider style; by 'traditional,' we meant there is an American tradition of cider."

Broad and bearing many versions, orchard-based ciders home in on two fundamental factors: the apple and the land it came from. The cider should speak to both and share the story from there.

Cider Apple Sage

Stephen Wood, Poverty Lane Orchards and Farnum Hill Ciders, Lebanon, New Hampshire

When it comes to traditional cider apples, not many people today have been growing the varieties in North America longer than Stephen Wood. Although his younger intentions were not to stay on that same farm, Wood has been working on Lebanon, New Hampshire's Poverty Lane Orchards since 1965, when he was 11 years old.

"I didn't think I would be growing apples when I grew up," Wood laughs, detailing how the orchard has changed over the course of his tenure, from when he was 18 and managing 90 acres to when he bought it himself in 1984. "I had the great luck of realizing what I actually do well is grow apples."

With dramatic shifts in the commercial fresh fruit market leaning more toward bright, bouncy, and beautiful culinary apples, Wood and his wife, Louisa Spencer, also had the good fortune of admitting their current farm wasn't going to work with these increasing demands. Around that time, the couple went to visit family in England, driving through the orchards of cider-centric Herefordshire, and the apple seemed to hit Wood right on the head.

Hundreds of grafting trials later back at Poverty Lane, they found they were able to grow the English bittersweets and bittersharps they fell for back in the cider

country of the United Kingdom. They began the process of ripping out aesthetically beautiful apple trees and replacing them with inedible cider fruit in 1989, planting trees of Ashton Bitter, Harry Masters' Jersey, Medaille d'Or, Yarlington Mill, and more. Six years later, Farnum Hill Ciders was a bonded and licensed maker of cider in New England.

As a tenured orchardist, Wood easily grasped the concept of growing this rare fruit on his land, but he credits his cider-making education to the wine industry, nodding specifically at Cornell University and New York's Finger Lakes wine appellation. As for the "style" of his ciders, he calls it simply American.

"When we first started making cider, I was planning to make something in a 'Somerset' farm style, then maybe something French, finishing the fermentation in the bottle," Wood recalls. "A few years after we started this whole thing, I'm still taking winemaking classes and we were sniffing through some of our stuff, and we realized that we were smelling aromas and tasting flavors in cider made from apples of these orchards that we never encountered in England and France. We had this epiphany of 'Why are we trying to make imitation cider? This is America, damn it! Let's just make something delicious; we don't have to adhere to tradition.' And that's what we did, and that hasn't changed."

The fruit was, has been, and is the driving force of Wood's production throughout the half-dozen ciders Farnum Hill regularly bottles. The ciders aspire to show the full expression of the fruit with a sense of their place, from single-varietal bottlings like the bittersharp Kingston Black to blends like the pleasantly effervescent Semi-Dry.

Although he's been growing longer than most producing cider on the continent, Wood says he would never call himself an expert. "I have strong notions about what I think is a high-quality cider apple, but other people have other notions about how to grow cider apples, and I want to learn from them," he says, respectfully naming other orchard-based cider producers across the country. "I think if our cider is any good, it's because we know how little we know. . . . It's possible to have been doing something longer than everybody else, it's even possible to know more than almost everybody, and still know nothing compared to what there is to know [about cider]. The tradition here is acknowledging our own confusion."

> 66
>
> We had this epiphany of 'Why are we trying to make imitation cider? This is America, damn it! Let's just make something delicious; we don't have to adhere to tradition.'"

For the Love of Apples

Autumn Stoscheck, Eve's Cidery, Van Etten, New York

Autumn Stoscheck loves apples — so much that at age 21, she took her life savings and started her own cidery in New York at the turn of the century.

"It was sort of fortuitous that, in 1999, Steve Wood [of Farnum Hill Ciders and Poverty Lane Orchards] was on the cover of *Fruit Growers News* and I read about his orchards, what he was doing, and it blew my mind away," Stoscheck recalls.

Shortly after, she found herself visiting Wood and his pedigreed orchard in Lebanon, New Hampshire. According to Stoscheck, Wood graciously took her under his wing and allowed her to take cider apple tree scion wood back home to the Finger Lakes.

From there, she dove in. She partnered with sixth-generation orchardist James Cummins and his 60-acre orchard to start Eve's Cidery in 2000, and today, the unlikely duo and Stoscheck's husband, Ezra Sherman, have been replanting an old farm purchased from her grandmother with organic cider apple trees to supply their production.

Times have changed since Stoscheck first started. She remembers much of the cider she tried when first launching Eve's was the result of funky, old rundown cider mills in rural Appalachia. "I had had some of that cider, and before I met Steve, it

had never occurred to me that there was something other than a country tradition," Stoscheck says. "Even at that time, James's orchard was a U-pick and sold fresh fruit. We completely changed our entire reason for being, which now exists for making the cider. It was sort of one of those evolutions that happens from just being in it, watching, realizing, and understanding what you're doing."

Stoscheck's deep affection for apples is a specific romance — she loves the apples that make great cider, apples of purpose.

"I think in the broadest sense, I would define a cider apple as an apple you're using intentionally," she says, noting that it takes time to grow an orchard with such purpose. "We're just learning what we like, what grows well here, what ferments into a cider that expresses what it is that we want to express and expresses what our piece of land is about. In that sense, that intentionality is going to take a couple more decades."

Getting explicit, Stoscheck says her true loves are European bittersweet apple varieties, those with levels of tannin, unique flavor, and distinct aromas that ferment into "some pretty complicated and wild flavors that are more toward the earthy end of the spectrum," she says. "To me, that's a defining characteristic of cider."

Her love might be focused on bitter-sweets but her orchards are not: Eve's estates grow bittersharps, aromatic, and heirloom varieties as well, the latter two being used for aromatics although they are not traditional European cider apples.

The majority of the cidery's production is done through the Champagne method, a process that requires season-by-season steps to create the "tiny, persistent bubbles" and a bright, creamy mouthfeel. For Stoscheck, it isn't always about the bubbles — Eve's also proudly produces two still, dry ciders of design and an ice cider made from cryoconcentration (the freezing of late-harvest juice). She thinks the expectation of cider having bubbles comes from its American heritage as a beer alternative, an assumption she is set on changing.

"I believe that if you're going to have bubbles in a cider, it's not an afterthought, it's not there because it's supposed to be bubbly," she says. "The truth of the matter is bubbles, in a way, are like sugar in the sense that you can start with nothing and have a little bit of something after you put bubbles in it. If you just have a dry, still cider, you actually have to be using great fruit and making a great product, having a really well-balanced cider that has the structure, that has enough going on that it's really exciting to drink."

Nearly in its second decade, Eve's Cidery is set on not only growing apples of intention but making ciders with a purpose to excite its drinkers.

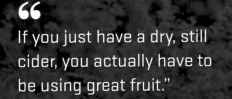

> 66
> If you just have a dry, still cider, you actually have to be using great fruit."

Orchard Heirloom

Kevin Zielinski, E.Z. Orchards, Salem, Oregon

E.Z. Orchards has long been revered for the vivid rainbows of farm-fresh produce bins and the sweet cider donuts fried in the window of its farm market, a vintage-design barn on the roadside of the Zielinski family's third-generation orchard. An incorporated farm in the south end of Oregon's heralded Willamette Valley since 1929, E.Z. Orchards has also been a quiet proponent of orchard-based, low-intervention, French-inspired cider production well before its ciders went commercial in 2009.

"At that time, people look at you like you're part kooky for wanting to do a non-inoculated ferment and not to use SO$_2$," orchardist and cidermaker Kevin Zielinski says. His ciders ferment naturally and are followed by the lengthy, aroma-driven *méthode ancestrale* — a production of sparkling wine that stops primary fermentation before it is complete and allows secondary fermentation to occur in the bottle without the addition of sugar, finishing when fully dry. "The idea is to transpose fruit into its most realized state without intervention."

Naturally, the orchard business brought Zielinski to cidermaking. He was making wine with a friend when another wine colleague reached out asking if he would propagate some French bittersweet scion cuttings for him. When a job took the colleague out of state, Zielinski was left with the apples and the curiosity to ferment. "I still see the apples when it's going into the bottle," he says. "I still recognize the fruit as it's coming off the tree, and I'm curious how it's going to develop because the juice is just a product of the apple to me, it's not the only thing happening."

Today, among other crops and culinary production, the Zielinskis grow ten varieties of traditional French bittersweet apples, plus four heirloom apple and four perry pear varieties. Zielinski divides the fruit between his two different intentions of cider and perry: French-style and North American. At the end of the day, his production is the same, but it comes down to the variety choices that make the ciders what they are.

"This is a cider that is about the fruit," he says, driving the focus that both styles have a sense of place — the North American showcasing domestic heirloom apples, and the French-style focusing on bittersweets seen in the land of the Franks. Both highlight their heritage. "It's different traditions, and this is what each culture would have on a daily basis."

Those traditions, Zielinski's tenacious production style, and his own family legacy help to build the foundation for E.Z. Orchards and their ciders. A zealot to his craft but a gentle advocate, Zielinski recognizes his dry ciders have entered a market where sweeter, fast-ferment ciders are the majority of what the public knows.

"The market is defined more by the modern cidermakers because there was very little memory of traditional cider styles in North America," he says, referencing the rapid growth of the United States cider market in its short window of time. "Expecting that the public may be interested in something that I was interpreting from a traditional sense wasn't really the reality. I'm not trying to compete so much with modern cidermakers, I'm trying to create a cider that sits on another part of the cider shelf that is also relevant."

Zielinski looks at the growth and future of the cider industry in comparison to that of wine, not so much in a production or agricultural manner but in following a pattern of public preference.

"As people experience cider, scholarship follows, and they start to go toward what expands their experience," he says, noting how as a palate matures, it gains appreciation for more artisanal offerings. "The Pinot Noir growers in the Willamette Valley experienced this. It was a struggle at first, and now they're world recognized for what they grow. . . . Things happen in decades, not in years."

Closing in on 90 decades of orcharding, E.Z. Orchards and Zielinski appreciate the notion of maturation with time, allowing their fruit and their ciders to grow up with the industry, cheerleading tradition and heritage along the way.

"

I still see the apples when the cider is going into the bottle."

Tasting Orchard-Based Ciders

TASTING ORCHARD-BASED CIDERS

Orchard-based ciders are all about the apple. They showcase the natural flavor of the apple varieties and the way the juice is fermented, rather than using flavorings added after the cider is fermented for supplemental complexity.

1. EXTRA DRY

Poverty Lane Orchards and Farnum Hill Ciders, Lebanon, NH

Bubbly and radiant in its bone-dry quality, this is maker Stephen Wood's top choice for sipping. In a complex balancing act, fruit, earth, acid, and tannin join forces to make a cider that could take the place of red wine at a meal. 8.5% ABV

2. WILD APPLE BLEND

Carr's Ciderhouse, Hadley, MA

Feral, unsprayed Massachusetts apples are foraged from wild trees surrounding this estate orchard, fermented naturally, and left still and sleek. A broad, fruit-forward and earthy mid-palate dries out while embracing natural tannins and a smack of untamed fruit. 6.9% ABV

3. RUSSET

Uncle John's Hard Cider, St. John's, MI

More than 30 different varieties of the russeted cultivar go into this cider, with Golden Russet as the dominant apple. Sharp fruit and earth are bounced off of golden apple and raisin flavors, with a finish that would match a semihard cow's milk cheese nicely. 6.5% ABV

4. ALBEE HILL

Eve's Cidery, Van Etten, NY

One of the two still ciders from this orchard-run cidery, this is a blend of heirloom and English cider apples. Aromas of citrus peel, baking spices, and fresh rain burst, while a palate of apricot, Key lime, and brisk red apple fill it out. 8.5% ABV

5. CRACK WILLOW

Twin Pines Orchards, Cider House, and Estate, Thedford, ON

Family farmed since 1968, these aged apples — like Northern Spy, Ida Red, and Golden Russet — go into this still apple wine. Leaning into its bright acid and rich fruit, the cider finishes clean, dry, and round. 12.8% ABV

6. SERIOUS CIDER

Foggy Ridge Cider, Dugspur, VA

English bittersweet apples meet their tart American heritage counterparts in this focused cider from cidermaking pioneer Diane Flynt. Crisp and dry with aromas of apricot, apple, and golden raisin, citrus flavors run through the center of the palate, with a streak of acid. 7% ABV

7. SPARK!

Alpenfire Cider, Port Townsend, WA
   Estate-grown traditional apples go into this bottle-conditioned blend. Aromas of bruised apple, biscuit, and raisin are met with full flavors, fine-grained supple tannins, and a semisweet finish. Try next to pork loin with crumbly blue cheese. 8.9% ABV

8. CIDRE

E.Z. Orchards, Salem, OR
   Traditional French-method production of naturally occurring yeasts fermenting the crushed apples and months-long maturation in the bottle result in flavors of dry apple and dusty earth, dazzling acid, and bittersweet tannins that embrace and release. 5.7% ABV

9. GRAVITY

Castle Hill Cider, Keswick, VA
   A blend of Virginian cultivars like Golden Grimes and English bittersweets like Dabinett, this still cider from the centuries-old property shows aromas of peach, white flowers, and citrus rind, with an off-dry palate that melds stone fruit with grapefruit flavors and acid. 7.1% ABV

10. OLD FANGLED HEIRLOOM BLEND

Whitewood Cider Co., Olympia, WA
   Urban cidermaker Dave White hand-selects a blend of Northwest vintage varieties, like Jonathan, Winesap, and Gravenstein, from an eastern Washington orchard for this coyly named cider. The result reveals round, orange citrus; bright acidity; and soft, crisp tannins. 6.7% ABV

MODERN CIDERS

For all the dreamers and the innovators, the bohemians and the experimentalists, and those who preach the unorthodox word and create new traditions, there is a big, wide world of modern cider just waiting to be discovered for both the producer and the drinker. Oftentimes a "gateway drug" into the broader category of cider, the term "modern" cider is an ambiguous, abstract one, so that interpretation can be anything. "Modern" cider can refer to the apples used or the adjuncts featured, like hops, hibiscus, or lemongrass, neatly packaged in both six-packs and 750-milliliter bottles and flowing from a bar tap near you.

THE HISTORY

The majority of modern consumers came face-to-face with cider by way of Angry Orchard — craft beer virtuoso Boston Beer Co.'s sister cider brand, kitted out in six-packs sporting an ornery tree icon with a semisweet, fruit-forward flavor profile. This was just 2011, and cider names like Woodchuck, Crispin, and ACE had long been on the circuit but not to the international recognition that Angry Orchard had received virtually overnight.

Unscrambling the legal and industry-wide muddle of what is or isn't cider, modern cider expositions are those without boundaries. No orchard? No press? No tanks? No problem. The dream of making cider is still alive and at one's disposal.

Custom production facilities are opening throughout North America, filling the holes where quality fruit and equipment are needed to make a cidermaking aspiration a reality. Wholesale juicers and high-quality concentrates are also more readily available from big fruit markets like Washington, Oregon, and Michigan.

THE BREAKDOWN

Largely built on culinary dessert apples and sometimes heirloom fruit (plus traditional cider fruit, when makers can get their hands on it), these ciders gush into the market with equally ambitious profiles, meeting the demands for consistent products, unique flavor combinations, and regular availability. This is the category where a consumer might find "made from concentrate" typed on the back label, with a major emphasis put on the packaging of that label. Electric and expressive wrappers line these cans and bottles, often seen with closures like pull tabs and bottle tops instead of corks and cages.

Targeting the craft beer drinker, modern ciders can be seen on a large commercial scale just as much as it can be made from someone's suburban garage. For other producers that are without orchards or access to consistent traditional fruit, this variation of America's original drink is a new sound for this oldie — but goodie — of cider.

Contemporary Design

Bruce Nissen, Jester & Judge and LDB Beverage, Stevenson, Washington

If modern American cider had a face, it might be that of Bruce Nissen's. Graced with a constant smile, Nissen is amiable yet pragmatic, and few people know modern cider — the product — in the United States like he does. He's chased the dream, starting Fox Barrel Cider in 2004 with his stepbrother-in-law. He's partnered with another to build their dream, launching Crispin Cider Co. three years later. He's made the move to corporate, selling the two brands to MillerCoors for one of the largest cider industry sales on record, and he's come back on his own again to craft with the opening of a co-packing beverage facility and cidery in Stevenson, Washington, making beverages for those who lack the production capacity.

When Nissen launched Fox Barrel in 2004 in Colfax, California, few makers on the West Coast were making cider with fresh-pressed juice, and that was a quality sacrifice he wasn't going to make. "Our theory was if we did something different, maybe we could light ourselves on fire and somebody would come to watch us burn," Nissen says. "And we did; we did fresh-pressed [juice] production when nobody was doing that out here."

Three years into the business without much revenue to prove it, Nissen says they realized they had way more equipment than they did product to keep busy with, and the concept of contract packing, or co-packing, came to light. The team began producing and packaging products on a client basis, from sodas to beer and cider, which led them to what would become Crispin Cider Co. — and the first craft acquisition for beer giant MillerCoors when it sold in 2012.

Nissen brings the co-packing mentality today to LDB Beverage, his colossal operation that sits 100 feet from the Columbia River on the Washington side. The state-of-the-art facility is home to Nissen's Jester & Judge brand and has the ability to produce anything from 150 to 50,000 cases on a daily basis, providing opportunities for producers, large and small.

"If someone comes to me and says, 'I want to make a cider,' I will sit down with them, road map it, and say, 'Think about these choices; now make one,'" he says of balancing creative aspirations and economic realities. "We get to be in this spot right now where at least once a week, someone comes into our facility with a dream. They don't know how to get their dream to a reality, so we help them. We do product actualization for people."

One of the main products on the line is Jester & Judge, driven by the motto of "seriously whimsical." The house brand produces ciders like the Marion Perry (a blend of Oregon marionberries and fermented pear juice), Pineapple Express (tropical pineapple-infused cider) and Columbia Belle (a cider melded with juicy peach and fresh mint). Each bottling grasps both the notion of contemporary cider and that of fresh-pressed juice, compromising little to create a consistent and easily consumable cider for the droves of modern cider drinkers.

Nissen says he sees cider in two different camps — "the sessionable," an everyday sipper that fills a pint glass, and "the celebratory," a stronger-structured cider that is sipped from stemware. He has much respect for both, but sees the ciders he makes falling into the first.

"I want to be your Tuesday-night cider, not your Friday-night, special cider," he says with a laugh. "My motto is, 'I make cider for America's fridge, not mine.'"

With experience on his side, Nissen views cider today as a blank slate. "Cider makes the sheet on which you paint," Nissen says coyly. "If you need it simple like Ansel Adams, God bless you. If you want to spice it up and be Pablo Picasso, I'm into that, too, as long as it works. I think there's a million things that can still be done that haven't been done."

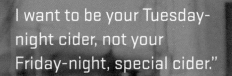

66
I want to be your Tuesday-night cider, not your Friday-night, special cider."

Washington's Darling

Joel VandenBrink, Seattle Cider Company, Seattle, Washington

Joel VandenBrink believes in the business of local cider. As the founder of Seattle Cider Company, Seattle's first cidery since Prohibition that became one of America's sixth-largest independently owned cideries in just two years, he understands the concepts of manufacturing and packaging without losing the essence of why he started a cidery in the first place — the fruit.

As an orchardless cider producer in the twenty-first century located just a couple miles south of downtown Seattle, VandenBrink's top priority was to get his hands on quality Washington apples from experienced Washington farmers. The next step was to get his beverage into as many hands and mouths as he could.

"I had always known I wanted Seattle Cider to be everyone's go-to, ubiquitous, gateway cider," VandenBrink says, with his cider now in more than a dozen states nationwide and also in half a dozen international markets. "I wanted to use all Washington-grown apples, not too sweet. For my flagship products, I wanted something a cider connoisseur could enjoy and a newbie could drink when they're getting into cider. I wanted something clean, good, and apple forward — not apple juice forward, but the apple itself."

Since commercial orchards aren't typically found within Seattle city limits, VandenBrink calls upon a select few orchards for his custom blends, pulling everything from heirloom Gravenstein and Esopus Spitzenburg to culinary favorites Granny Smith and Gala. The cidery also works with the Washington State University Research and Extension Center an hour north of the city that specializes in cider varieties and tree fruit favorable to the Pacific Northwest climate.

According to the beverage entrepreneur, VandenBrink says Seattle Cider came together as a "perfect storm." After starting Two Beer Brewing Co. in Seattle in 2007, he was contracted by a brewery to make their beer when, fortuitously, the space behind his production facility opened up and allowed him to add more tanks to custom-make this beer. That gig didn't end up lasting, but the space and equipment did, and VandenBrink saw an opportunity — a former cellar assistant of the brewery suggested they start a cidery and that he could run production. In the same year, VandenBrink was diagnosed with Crohn's disease, and with that came a lifestyle reduced in gluten, sugar, and lactose. All signs were pointing to cider.

Only 16 months after launching, the cidery was producing as much as the mature brewery. Since size can matter, the label of "craft" in regards to cider is something VandenBrink says he identifies with.

"The term 'craft cider' is still a very abstract term, or 'cider' for that matter — even the government doesn't know what it is," he says. "I didn't want to make the Budweiser of cider. In order to set ourselves apart as a craft cider producer, we never use concentrate, artificial colors, or high-fructose corn syrup, and I wanted to set our price a couple bucks higher so that the consumer would trade up."

In a shrewd marketing move, Seattle Cider's retail cans and bottles offer more than the apple-forward beverage — they offer an education to the consumer on the label by listing the exclusivity of Washington apples as the source and even a Brix scale to note sweetness. "In my mind, this started to educate the consumer to realize there's a difference," he says. "We have what not many others on the shelf can claim — we're truly local. All of the orchards we use are within 120 miles of the cidery, and the juice gets to us within two hours of being pressed. A lot of people know Washington apples, and that's enough in its own for people to sign up with it."

> "
> We're truly local. All of the orchards we use are within 120 miles of the cidery, and the juice gets to us within two hours of being pressed."

From Boston Lager to American Cider

Ryan Burk, Angry Orchard Hard Cider, Cincinnati, Ohio and Walden, New York

When Jim Koch chanced upon his great-great-grandfather's lager recipe in his father's attic, cider wasn't even a gleam in his eye. It was 1984, and the craft beer movement was on the up-and-up, a charge led largely by Koch and his Samuel Adams Boston lager. In 1997, Koch and Boston Beer Company found themselves on another frontier, pioneering permeation into the fermented apple industry with HardCore Cider. The cider quietly sat in the New England market, and after a 15-year presence, it slid over to make way for what arguably landed modern cider in the American spotlight — Angry Orchard Hard Cider.

As the largest independently owned cider brand in the United States, Angry Orchard can take credit for nearly 60 percent of cider sales in the country, according to recent IRI data.

Drawing influences from English cidermaking with traditional cider fruit, head cidermaker Ryan Burk says the history of the company and its focus on high-quality materials is the base of what has made Angry Orchard the most popular cider in the country.

"I think it could have been anybody, but our attention to quality has brought people to the fold that may have not otherwise," Burk says. "That's a quality statement from drinkers to us. . . . For a cidermaker, you can't really get any better than that."

Burk welcomes questions about Angry Orchard's apple base: its juice, its newly acquired 150-year-old orchard in upstate New York, and its use of apple juice concentrate — the processed, storable, concentrated juice that allows for rehydration and fermentation. His answer is shared with respect for the product.

"We go to great lengths and always have to source traditional cider fruit. For instance, people mention the bigger producers who use concentrate — which we do, sometimes, but not all the time — are sourcing low-value product," Burk says, noting that Angry Orchard's size and success set them up to be an easy target. "Although we do travel outside of America for product, we go to Normandy and Brittany, two of the greatest cidermaking regions in the world, to source our bittersweet fruit. We don't have to do that, we *choose* to do that, and that's based on our tradition of cidermaking. We also look to England for some of that fruit. We're really proud of that sourcing, and I think that's what separates us."

With the main headquarters for large-scale fermentation staying in Cincinnati, Burk's heart and passion is in the 60-acre orchard and cider house in Walden, New

York. A native of the Empire State, Burk is able to grow his own fruit on the land, develop a barrel-aging program, and experiment with wild yeast cultures, "an expression of self."

His focus is on the apples — in raw fruit or concentrate form — and the blending of ideal varieties for Angry Orchard ciders, from the core grocery line to the higher-end Orchard's Edge and Cider House Collection.

"If you took blending out of the art, I would be less interested. I want to take 20 things and make one cider out of it, or seven ciders out of it," he says. "That's what's exciting, and that's what we do every step of the way."

Even with 20-plus years of experience for Boston Beer's cider production, Burk agrees with many that cider is still yet to be defined in North America, and that starts with placing importance on the apple. Drinking cider "is an agricultural act; we're adding value to the act," he says. "That's what matters more than cidermaking. It's the farmer, the grower, the history of the fruit. Cider has the opportunity to define itself; the future is in the apples. Once we start to define that, we'll be able to see what American cider is."

> "
> If you took blending out of the art, I would be less interested. I want to take 20 things and make one cider out of it, or seven ciders out of it."

Tasting Modern Ciders

TASTING MODERN CIDERS

Whether made from concentrate or fresh-pressed juice, modern ciders encompass a variety of flavors and production methods.

1. DRY

Seattle Cider Company, Seattle, WA

One of the first truly dry ciders to hit a large-scale commercial market, this unfiltered cider is canned in a tallboy in four-packs and highlights pure, clean Washington dessert fruit. Flower, orange, and green apple aromatics supply the bouquet while stone fruit and red apple fill the dry palate. 6.5% ABV

2. GREEN APPLE

Angry Orchard Hard Cider, Cincinnati, OH

Featuring culinary fruit sourced from the Pacific Northwest, this cider is tart and tangy as one might suspect, showing Granny Smith apples the most. Off-dry and puckering, the fruit-forward palate is crisper and brighter than other offerings from this cider megastar. 5% ABV

3. CIDERKIN

Argus Cidery Fermentables, Austin, TX

A segue from the more traditional ciders of the Argus Cidery, this line of cans delivers straightforward cider sipping. Made from reconstituted apple pomace and fermented naturally, this low-alcohol cider is dry, fruit-forward, and incredibly refreshing. 4.5% ABV

4. THE HATCHET

Sonoma Cider, Healdsburg, CA

This father-son-run operation uses organic Northwest apples to make their Sonoma-based ciders, with sleek, self-explanatory packaging and amiable flavors. The Hatchet is off-dry, uncomplicated in its fruit, finishing crisp and true. 6% ABV

5. ORIGINAL

Apple Outlaw, Jacksonville, OR

Producing cider from mostly estate fruit in southern Oregon, the profile of the cidery is one that adheres to its environment — creative and fruit-forward. The flagship product is just that, with semisweet fruit and racy acidity. 5.5% ABV

6. ACE JOKER

California Cider Company, Sebastopol, CA

The driest of the flavor-enhanced lineup from this family-owned cider house, this approachable cider is easy drinking, full of juicy fruit and simple acid. Let the higher alcohol content be known and try pairing it with grilled citrus salmon. 6.9% ABV

7. THE SAINT

Crispin Cider Co., Colfax, CA

Fresh-pressed Washington and Oregon apples go into this unfiltered cider, fermented with Belgian Trappist yeast and sweetened with pure maple syrup. The aromatics are toasty with notes of maple and baked apple, and the palate sips like ripe apples, tropical fruit, and baking spices. 6.9% ABV

9. FIRST PRESS

Ciderboys Hard Cider, Stevens Point, WI

Focused on flavored ciders at affordable, approachable pricing, Ciderboys also makes a traditional cider of apple and only apple. Loaded in fruit, flavors of clean apple are matched with crisp acid and cravings for cheesy puffs. 5% ABV

8. DRAFT

Bold Rock Hard Cider, Wintergreen, VA, and Mills River, NC

Two production facilities spanning two states, this cidery needed to double to feed the hunger for its ciders. The Draft represents both states of Virginia and North Carolina, an amber cider of smooth apple fruit, friendly acid, and an off-dry finish. 4.7% ABV

10. WOODLANDER

Grizzly Ciderworks, Milton-Freewater, OR

Two college buddies started a cidery, without apples and hardly with equipment. A number of variations and rotating seasonal selections later, the original ciders still remain, like this Belgian wheat beer–influenced sipper of fresh fruit, spices, and citrus. 6.7% ABV

SINGLE-VARIETAL CIDERS

Cidermaking is both an art and a science, and most cidermakers find themselves straddling the line between the two. Many believe that blending — harmoniously composing a mix of apple varieties — is the true art, while a single-varietal showcase is arguably the most naked a cider can be. Standing alone, single-varietal ciders are some of the most transparent and revealing routes to expressing the apple.

Grandiloquent prose aside, single varietals certainly do one thing: bear everything, scars and all, that particular fruit has.

THE HISTORY

Most East Coast growers will talk about Thomas Jefferson's favorite apple varieties — Albemarle (Newtown) Pippin, (Virginia) Hewes Crab, or Esopus Spitzenburg, depending on who and where you ask the question. Many of the apple varieties originally cultivated on his Monticello estate and South Orchard in Charlottesville, Virginia, sprouted from those grounds and have carried their single-varietal success to the modern era of cidermaking.

Kingston Black, the regal bittersharp of Britain, is one of the more long-standing single-variety bottlings in its home of the English West Country. Reason being, it has so much complexity on its own — tannin, acid, and sugar — without blending. American heritage apples like different Pippins and Russets, Gravenstein, and Arkansas Black also have proven their worth throughout the decades as single-variety triumphs.

THE BREAKDOWN

Much like pure varietal showcases in wine, apples used in single-variety ciders need to be able to stand on their own, and some do better than others. Exemplifying the characteristics of the fruit, the ciders that sport the varietal name typically have the right chemical composition — pH, tartaric acid, sugar levels, and so on. Where they might fail as popular culinary or dessert apples in the fresh market, these apples are best when they are fermented — blended or alone.

Throughout North America, a larger number of cideries produce single-varietal offerings in smaller numbers and are often doled out to specific markets.

For Field Maloney of West County Cider, the United States' oldest commercial cidery since Prohibition, single-varietal ciders were and are the foundation of his production. Maloney says he'll always make single varieties, like his father did when he started in 1984. He says part of the appeal of making the cider is the "study to see how we can distill the essence" of the specific fruit from which it comes.

66

It's indeed bad to eat apples, it's better to turn them all into cider."

— Benjamin Franklin

Historical Revival

Chuck Shelton, Albemarle CiderWorks and Vintage Virginia Apples, North Garden, Virginia

The late, celebrated author Michael Crichton once said, "If you don't know history, then you don't know anything." This sentiment is something Chuck Shelton of Vintage Virginia Apples and Albemarle CiderWorks can agree with — the family farm has built its business on reviving historical apples from the commonwealth.

Deeply rooted in Virginia agriculture, Shelton and his siblings grew up around apples. When his parents decided they wanted to embark on a homesteading "retirement project," the whole family went in on purchasing a small farm in southern Albemarle County. Today, the Shelton's hand-selected Rural Ridge orchard consists of more than 250 different cultivars, including varieties historians haven't seen in a commercial sense for more than a century.

"After Prohibition, the orchards were gone, they disappeared," Shelton says. "We only have a few of those varieties left, so we're trying to build that back, but it takes a lot of time to grow apples trees. It used to take almost a lifetime to get an orchard back."

At least in his lifetime, Shelton will see his vintage apples thrive — and in less than two decades. After attending apple tastings at Thomas Jefferson's historic Monticello estate in Charlottesville, Virginia, in the mid-1990s, Shelton and his sister and business partner, Charlotte, were drawn back to the fruit of their childhood. Led by apple expert, historian, and scion Tom Burford, the siblings tasted from 50 different heirloom apple varieties that they didn't even know existed.

Charlotte (the "historian" of the two, according to her brother) was compelled to bring these varieties back to their family farm. Together, they selected 20 varieties to plant. By the early 2000s, they had upwards of 200 different apples in the ground.

"It's a small orchard, but it's pretty eclectic in terms of its variety," Shelton says. "We had access to apples that were hard to access, so we started grafting and selling trees. . . . That's the base of our business."

The cider production of Albemarle CiderWorks was also rooted in history. "The idea of cider came from what we were going to do with these apples," Shelton says with a smile. "If you produce in good quality, you can sell some of them. If you don't, you've got to do something with them. That draws you right to cider, which was the original purpose of the apple in America. . . . The way of preserving fruit, for something wholesome from the fruit."

For Shelton and his cidermaking, he calls upon his vintage American apples to

> " After Prohibition, the orchards were gone. . . . We only have a few of those varieties left, so we're trying to build that back, but it takes a lot of time to grow apples trees."

make his "American cider" rather than traditional European and French cider fruit. "I don't actually need cider apples, but I do think it brings a new dimension," he says, adding that he doesn't think the drink is defined by cider apples. "In fact, there was a lot of cider made in the United States by apples that originated here and was pretty good stuff."

With American history as his foundation, Shelton is making cider from apples of lengthy pedigree, like Albemarle (Newtown) Pippin, Virginia Hewes crabapple and Winesap seedlings, Arkansas Black, and Black Twig. "I've decided to make good cider for me or for someone else in the past," he says. "Newtown Pippin, it's one of Jefferson's favorites. George Washington used it; it is a 300-and-some-year-old apple. . . . We have a couple of letters from Jefferson where he ordered a thousand gallons of it."

Name-dropping apple varieties is something of a trademark for Albemarle; single-varietal ciders are the main focus for Shelton's production. From the old apples to new plantings of recent cultivars like GoldRush, the apples that get their name on the label are not determined by a tested scientific hypothesis. Shelton says he chooses them the same way he and his sister chose to put those apples in their ground — by taste.

"There are other apples out there; you just have to hunt around for them," he says. "When you find one that has the acidity, the sugar levels that give you the good alcohol — generally nice pH that you need for fermentation — then you're looking for the acid bite. To me, that's what makes the cider."

Second-Generation Bushwhacker

Field Maloney, West County Cider, Colrain, Massachusetts

Field Maloney remembers which blocks of apple varieties were planted on his family orchard by what grade he was in when they went into the ground. "We cleared the land for our first block of Golden Russets back in 1986; I was in tenth grade then. In eleventh grade, another block of the orchard was planted," Maloney chuckles.

The first commercial cidery in the modern United States, West County Cider began its single-varietal-focused production in Colrain, Massachusetts, in 1984, hand-crafted by Field Maloney's parents, Terry and Judith. He was in eighth grade then.

West Coast transplants, Terry and Judith were at the forefront of the Napa Valley home-winemaking scene, steered by their drive to create something from the land. After visiting with a friend in rural western Massachusetts, they thought they would plant their roots there — attempting true homesteading in the early 1970s by purchasing a 50-acre lot of woods and clearing it themselves for an apple orchard.

"It's not the most auspicious place for an orchard. It's a very rough and forbidding spot to farm; steep, hilly, cold," Field Maloney says. "We cut down enough trees to build a house, for a garden, and for the first orchard."

The family was fervid about fermentation, high from the rush of California winemaking and the chance to make their

land sustainable with the old New England tradition of cider. They located a handful of unruly orchards, then groomed and picked them, producing their first hobby ciders in 1972. The three would bottle the ciders in the cellar of the house, where they still bottle today.

Twelve years after they first started making cider at home, West County Cider received its license from the state and was officially a bonded winery.

"My parents focused on taking this old rural drink and wanted to see what could happen if they applied modern scientific winemaking principles," Maloney says, explaining that most of the production was and is single varietal, like wine. "Being steeped in the past, knowing this was a major, big-time drink in American culture that totally vanished off the face of the earth . . . they thought, 'Let's bring it back.'"

Influenced by the rich history of the region and inspired by the area's farmers who still made small, home-ferment batches for pleasure, the Maloneys helped launch CiderDays, a now annual community celebration "of all things apple" in Franklin County, Massachusetts. In 1994 when the festival started, it was to shine a light on cider by showcasing the hobbyists, the growers, and the apple itself.

"Right around 2005, we'd have CiderDays, and all the cidermakers in the country would come. That's how small cider was then," Maloney remembers. "If you wanted to make cider back then, you had to believe in it and had to use very creative ways to make it on no budget at all. It was a lot of Yankee ingenuity."

For the Maloneys and those who started making cider before the boom, their instincts on the drink have been validated. Maloney laughs, reminiscing how his late father, Terry, would have loved to see where cider is now. "For us, we spent two of our three decades crying in the woods, saying, 'Drink cider!' By sticking with it, gradually getting people to open up, we're showing our rural culture. It was the main drink of our country, then it vanished. . . . Now cider is on everyone's focus."

With the single-varietal cider focus, the Maloneys treat the original orchard as a "test" block for different apple varieties, allowing them to narrow in on what grows best on the land. Decades and five planted acres later, and with the acquisition of centenarian orchards, they have been able to see what each apple can do.

Maloney doesn't know if single-varietal ciders explain the whole process of creating the beverage, but admits the style is quite naked and exposed. "You're getting a very simple and clear expression of what it is," he says. "It's hard, because it depends on the deftness of the hand of the cidermaker, the quality of the fruit, and all the billions of things that go into what makes a finished bottle. But that's why I'll never give up on single varietals."

And he isn't giving up on cider in general, even when others label it as a trend. "Cider has a place at the table. It's not a passing fad, because it makes so much sense," Maloney says. "Apples, even though they're a European export like to some degree the country itself, they're the closest thing to our home fruit. . . . I think that cider is only going to get better and better over time. The apple is suited for growing in so many places in America, the possibilities are just getting going."

66
My parents focused on taking this old rural drink and wanted to see what could happen if they applied modern scientific winemaking principles."

Tasting Single-Varietal Ciders

TASTING SINGLE-VARIETAL CIDERS

Single-varietal ciders rely on apple varieties with a flavor profile that is complex enough to stand on its own — just the right balance of acidity, tartness, and sweetness.

1. INCLINADO SIDRA

Tilted Shed Ciderworks, Windsor, CA

This small-batch, limited-release Gravenstein cider is Spanish *sidra*–inspired and split into two releases, half still, half sparkling. The still cider is asking to be long poured to release the funky citrus aromas and tannic flavors, while the bubbles in the sparkling version showcase the aromatic profile on first sniff. 7.5% ABV

2. DRY BALDWIN

West County Cider, Colrain, MA

Featuring a variety that was first discovered in Massachusetts in the early 1800s, family-run West County bottles its best seller completely dry, round, and smooth in apple fruit. Juicy yet dry, the cider is true to the 100-year-old trees it is picked from. 6.7% ABV

3. SPARTAN DRY

Montana CiderWorks, Darby, MT

Savory aromas and flavors of crunchy pear are prevalent in this apple that comes from Montana's Bitterroot Valley. Exhibiting white wine–like flavors of dazzling acid, baking spices, and clarity in fruit. Lovely with buttered popcorn or grilled whitefish. 5.5% ABV

4. KINGSTON BLACK RESERVE

Poverty Lane Orchards and Farnum Hill Ciders, Lebanon, NH

A showcase of a single vintage from this orchard's Kingston Black crop, the bitter-sharp variety comes through in spades of tannin and acid, with dusty fruit and apple blossoms filling out the still and vastly complex cider. 8.5% ABV

5. DRYHOUSE JONAGOLD

Embark Craft Ciderworks, Williamson, NY

Fermented from a specific higher sugar and acid strain of Jonagold by the fifth-generation New York apple grower, this cider is sessionable but far from simple. The tipple boasts great structure of clean acid and freshly harvested apple fruit. 7.4% ABV

6. GOLDEN RUSSET

Wandering Aengus Ciderworks, Salem, OR

Considered as the best flavored of the Russet varieties, the apples for this cider hail from the cidery's Salem hometown as well as Ashland in southern Oregon. Dry with golden, honeyed aromas, the acid is brilliant and finishes bright. 9% ABV

7. CIDRE MOUSSEUX SEC MÉTHODE TRADITIONNELLE

Cidrerie Michel Jodoin, Rougemont, QC

The dry rendition of this sparkling McIntosh cider from a third-generation orchardist and cidermaker is radiant and bouncy, with endless bubbles, and full of fruit, floral flavors, and bready yeast. Although vintage specific, the cider is consistent and demands triple-cream cheese. 7% ABV

8. VIRGINIA HEWES CRAB

Albemarle CiderWorks,
North Garden, VA

A crabapple of many names; this tenured orchard knows them all and deemed the variety, rarely seen by itself, worthy of its own cider. Big but balanced in body and flavor, floral aromas are packed with high-acid tang and robust profile in this complex sipper. 10% ABV

9. PIPPIN

Dragon's Head Cider, Vashon, WA

Ambiguously named for the many Pippins growing on trees through North America, this is exclusively Albemarle (Newtown) Pippin, largely from the estate orchard of this Seattle-area island. With aromas of fresh green herbs, tangy apple, and soft pear, some moderate farmhouse funk spills out, and astringency and fruit dance in the finish. 6.9% ABV

10. NORTHERN SPY

Citizen Cider, Burlington, VT

From the happening cider town of Middlebury comes Citizen Cider, a modern maker with traditional inspirations. This single-varietal sipper is based on a New York cultivar that has creds for great pie, and the succulent, sumptuous cider wisely turns up mandatory acid in its full-bodied fruit-forward finish. 6.9% ABV

DESSERT CIDERS

Liquid gold, silky, and sweet, dessert ciders can vary in style but include only a few categories in North America — ice cider, late-harvest cider, and pommeau (an apple brandy–fortified style). This and apple eau-de-vie, or apple brandy, both have French origins, so it only makes sense that ice cider, North America's original dessert cider, also has a French connection in its Québec roots. Ice cider, or *cidre de glace,* might not have the centuries-old legacy that alcoholic apple dessert sippers like Normandy's Calvados or Brittany's Lambig have — nor is it a distilled spirit or fortified with one — but it is a modern French-Canadian invention that has caught on and lit up not just the cider industry but the wine industry as well. In addition to ice cider, the breadth of dessert cider options range from late-harvest fermentations to brandy infusions, allowing drinkers to mix, match, and experiment in pairings, preferences, and varieties.

THE HISTORY

Ice cider. Born out of hard times in the Québécois beverage world, ice cider has been embraced as the provincial drink. After the short-lived prohibition in the 1920s, cider slipped through the cracks of the new government-approved alcohol list, while wine, beer, and spirits were able to return to filling the glasses of the Canadian people. An ill-fated cider boom in the 1980s nearly crashed and burned the market, but

in 1990, French transplant and winemaker Christian Barthomeuf began experimenting with a similar recipe to ice wine, with the substitution of apples. He teamed up with fellow ice wine enthusiast and maker François Pouliot, and the two wrote the formula for ice cider, based off of the like-minded principles of ice wine: using freezing temperatures to naturally concentrate the apples, and fermenting into an extract of high sugar and intense flavors.

In 1995, Pouliot's Neige ice cider, under his La Face Cachée de la Pomme cidery (now called Domaine Neige), was the first commercial ice cider to be released, creating a new category for the beverage and filling a much-needed North American dessert cider niche. Although most ice cider is still, several cideries also produce a sparkling rendition of the drink.

Pommeau. The origin story behind pommeau, a blend of unfermented apple juice and brandy (traditionally Calvados, an apple brandy from Normandy), is a little less clear. Technically classified as a "mistelle," which is the combination of fruit juice and brandy, French legend has it that pommeau came from accidentally spilling fresh juice into a barrel of brandy. Romantic but unlikely given the tightly sealed quarters of a barrel, more plausible testimonies deduce that the addition of the brandy was to preserve the juice or that the fresh juice was added to alleviate the heat from the heady brandy.

Pommeau in France struggled with a similar ban like cider did in Québec, except that it spanned nearly four decades and was only outlawed to protect wine production. Back with authority (pommeau gained its official designation from the French government in 1991), the unctuous hybrid is making a comeback throughout North America as well.

Late harvest. Typically plucked from the cold fall trees earlier than ice cider but later than the regular harvest, these sugar-soaked apples are ready to produce a sweeter style of sipper naturally. Often still, late-harvest ciders are fermented in smaller, select lots — similar in fashion to their dry brethren — and provide a way for cidermakers in climates inhospitable to ice cider production to make a dessert cider.

Other dessert-style ciders are based on the pommeau-style or late-harvest model: blending with juice and apple brandy or picking fruit late in the harvest season when it is ripe, fat, and happy, then pressing and fermenting.

THE BREAKDOWN

Tumbling from a graceful, slender bottle like a liquid syrup interpretation of Rapunzel's hair down the tower, ice cider is almost as pure as the famous fairy tale. The fermented apple juice is produced in one of two ways. The first, and most common, is cryoconcentration. With this method, fruits are harvested even later in the season than late-harvest picking (usually by the end of October) and kept in fresh, cold storage until late December, when the fruit

is pressed and the juice freezes into a concentrated elixir in the naturally chilling temperatures of winter, usually undergoing fermentation the following month. Another method, cryoextraction, is more akin to the traditional techniques used for ice wine: the apples are left on the trees to freeze until late winter, are picked by the end of January, and are pressed and cold-fermented for a prolonged period of time.

The production of pommeau takes different steps depending on location and licensing, but most regions in North America require the apple brandy to come from a distillery (either in-house or out). The cidery's fresh cider apple juice can then be blended into the brandy — roughly one-third brandy to two-thirds juice — and aged in oak. The French require at least 14 months in barrel, but North American standards have not been set. The finished product — a crossbreed of the realms of cider and spirits — is typically between 16 and 18 percent ABV and is typically consumed as an aperitif. The sipper sounds and smells like a cider but talks and tastes like a spirit: high-octane apple aromas and flavors are met with a depth brought on by caramel oak notes and a mahogany tint.

There are more than 50 different producers of ice cider in Québec (many of them in the Montérégie), and courtesy of Québec's government-run chain of liquor stores, their product is widely distributed throughout the province. In 2014, Québec ice cider was legitimized as a protected and reserved appellation, and other colder regions are dipping their toes in the water. Eden Specialty Ciders in Vermont, a southern neighbor to the ice cider motherland, has also branded itself on its ice cider production, working with New England heirloom and European bittersweet varieties for their highly concentrated juice, while cidermakers like Spirit Tree and County Cider in Ontario use local varieties like Ida Red and Northern Spy to withstand the cold weather and condensed liquid.

North American interest is sprouting in pommeau-style and late-harvest sippers as well, with producers like Westcott Bay Cider and Tieton Cider Works in Washington State, ÆppelTreow in Wisconsin, Uncle John's Hard Cider in Michigan, South Hill Cider in New York, Carr's Ciderhouse in Massachusetts, and Merridale Estate Cidery in British Columbia.

> 66
> No fruit is more to our English taste than the Apple.
> Let the Frenchman have his Pear, the Italian have his
> Fig, the Jamaican may retain his farinaceous Banana,
> and the Malay his Durian, but for us the Apple."
> — Edward A. Bunyard, *The Anatomy of Dessert*, 1929

The Godfather of Ice Cider

François Pouliot, Domaine Neige, Hemmingford, Québec

Before François Pouliot was bottling the world's first commercial ice cider, he was producing music videos in Québec for Canadian sweetheart Céline Dion, singer-songwriter Daniel Bélanger, and '80s hairband Les BB. Four feature films are also under his belt, along with a string of awards and accreditations, but by the time he was 28 years old, Pouliot longed for something more.

He began making ice wine on his grand-parents' farm in Montérégie, but when they died, the young entrepreneur was outpriced on the sale of the property and needed to look elsewhere to continue his hobby. He found an old orchard with a beautiful, rustic stone home in Hemmingford, only 45 min-utes south of Montréal, and pulled the trig-ger, readying himself to plant a vineyard.

"There was a lot of work to do," Pouliot admits. "Then fall comes and we have all these apples, so we think, 'Why not try the ice wine with the apple?'"

With the guidance of his mentor and ice cider recipe creator Christian Barthomeuf, the two pioneers developed the technique of cryoconcentration in 1995. Thanks to Québec's cold winters and sur-plus of McIntosh apples, Pouliot dropped wine completely and picked up his novel creation. "Québec is an apple country, and

this is really a reflection of my terroir," he recalls thinking. "Why don't I stick to the apple and never mind the grape? We'll call it ice cider instead of ice wine."

Under the original handle of La Face Cachée de la Pomme (loosely meaning "the hidden side of the apple"), Pouliot's cidery bottled Neige Première, the world's first ice cider. In 1998, the cidery coined the term "ice cider" on its labels; later that year, *Journal de Montréal* published "ice cider" and "cryoconcentration" in a feature story for the first time.

"Most everybody works with that tech-nique," Pouliot says of cryoconcentration. "We work mostly with the McIntosh apple, and what's the signature of that method is the freshness — you feel like you are biting a fresh apple. Although it's quite sweet, because of the acidity of the apple, it's so fresh."

Pouliot believes cryoextraction, the alternative method, changes the profile and purity of the apple, compromising the integrity of the fruit overall. "The apple is totally different — it's more like exotic fruit of lychee, apricot," he notes. "If you didn't know it was apple, you could think it was an ice wine."

Shortly after the release of Neige Première, Pouliot found himself interpreting his ice cider to the countries that had begun to import it, including France. "The French thought, 'Oh, this is not cider, there are no bubbles.' And we said, 'It is cider, it's just apple juice fermented,'" Pouliot recalls, joking that *he* had to explain cider to the French. "There are bubbles in cider because you provoke it. You don't need to have the bubbles in cider; if you have bubbles in cider, it's because you decided to have bubbles."

Today, with 80 acres of orchard, distribution in nearly 25 countries, and dozens of ice cider producers following the path he paved, Pouliot can still proudly say he started a movement.

Now, he says there are ice cider producers everywhere, from Vermont and Michigan to Denmark, Sweden, and Germany. "They all came by here to see how we were doing it and were inspired by us. The ice cider was super premium. It brought a lot of people into the category, and they would discover other ciders. With ice cider, we can open doors."

66

Why don't I stick to the apple and never mind the grape? We'll call it ice cider instead of ice wine."

Inherited Cider

Kristen Needham, Sea Cider Farm and Ciderhouse, Saanichton, British Columbia

Sheltered by a forest with airflow gusting off the waters of the Haro Strait onto Vancouver Island, the original orchard for Sea Cider Farm and Ciderhouse was a third-generation handoff to owner Kristen Needham when she was a teenager. At the time, she was attending boarding school in Wales, drinking plenty of cider — paired with plenty of studying, she assures — when she found she had inherited a family orchard on British Columbia's Saanich Peninsula.

Life gave her apples, and she decided to make cider. "I was exposed to cider and the whole European culture around food and beverage at such a young age," Needham says. "When we decided to build the Ciderhouse, I wanted the experience for people to be one that was more in keeping with that tradition of sitting with friends and family, enjoying the experience of good food, good company. I thought that was central to Sea Cider. I wanted a business that really wrapped the whole experience around cider."

Along with the family orchard, Needham planted another 10-acre property with cider apples in 2004 and put that sentiment in action — her own ciders released in tandem with the opening of the estate Ciderhouse Tasting Room in 2007. The Ciderhouse is a large portion of Sea

Cider and Needham's personality, serving as a conduit for patrons to fully engage in cider. "Cider is for celebrating, and cider is worth celebrating," she states. "It has a place at the table."

A descendant of longtime ranchers and farmers in British Columbia, Needham says starting the Ciderhouse wasn't exactly stepping out of her comfort zone but more like coming into her own. "Without my family, I wouldn't have been in the business of growing apples and making cider," she observes. "Because cider has traditionally been an agrarian beverage and that's very closely tied to family, we are proudly family owned but quite community minded."

Today, Sea Cider Farm and Ciderhouse consists of three different orchard properties and collaborates with others in the area, with more than 60 varieties of bittersweet and bittersharp cider apples and heritage fruit. And all certified organic — another trait of Sea Cider's disposition. The farm is in compliance with Canada's National Organic Standards through the Pacific Agricultural Certification Society (PACS), a feat that requires minimal intervention and prevention of plant disease and infestations through soil biodiversity and general health.

Like in organic and sustainable standards across North America, Needham says she finds the lack of legal definitions in the cider industry to be a hurdle. "There isn't one definition for cider or one definition for what makes a good cider," she says. "This is a challenge for producers to raise awareness and show what cider can really be."

Part of Needham's cider revival is showcased through her "sticky" series, a line of four dessert ciders that aren't typically made in her northwestern corner of the continent. The list ranges from a pommeau-style cider made by fortifying fermented Snow apples — the original variety grown in Normandy and used in the classic French production of pommeau — to the Pomona, a still ice cider produced by pressing late-harvest crabapples and freezing the concentrated juice.

"All four speak to a tradition and give Sea Cider's take on that tradition," she says. "I think they show consumers some of the more unusual ways that one can create a cider out of apples. They speak to the diversity and the breadth of the cider category overall. I think for so many years, cider was one-dimensional, and we are now showing the world that that's not the case."

66

There isn't one definition for what makes a good cider."

Tasting Dessert Ciders

eden
VERMONT ICE CIDER

HEIRLOOM
BLEND

POMMEAU
19% alc   375 ml

SOUTH HILL CIDER
ITHACA NEW YORK

SEA CIDER
FARM & CIDERHOUSE

Pomona

Inspired by the Roman Goddess of Apples, Pomona is a concentrated elixir crafted by freezing then slowly fermenting farm-pressed British Columbian crabapple juice, yielding a still dessert cider with stone fruit and confectionary notes. Raise a glass of Sea Cider to the Apple Goddess!

17% alc./vol. 375 mL

2 TOWNS CIDERHOUSE.

POMMEAU
Barrel Aged Apple Wine | Batch 21
375 ML   19% Alc. Vol.

NEIGE
CIDRE DE GLACE – ICE CIDER
PREMIÈRE

6

8

9

10

7

Island, Washington

ott Bay Cider

nmeau

produced and bottled by
Westcott Bay Cider
an Island, Washington

16% ALC/VOL

TIETON
CIDER WORKS
Washington State

TIETON FROST
CIDER

ALC. 11% BY VOL

BIG B's CIDERS
Bourbon Barrel Aged
POMMEAU
Hand Crafted in Colorado

USDA
ORGANIC

ORGANIC APPLE WINE BLENDED WITH SWEET APPLE CIDER
Year: 2015   Batch: 1   Barrel: 1

                                            300ML
PRODUCED & BOTTLED BY:
NORTH FORK CELLARS LLC, Hotchkiss, CO
CERTIFIED ORGANIC BY THE CO DEPT OF AGRICULTURE

14% ALC/VOL

GOVERNMENT WARNING: (1) ACCORDING TO THE SURGEON GENERAL, WOMEN SHOULD
NOT DRINK ALCOHOLIC BEVERAGES DURING PREGNANCY BECAUSE OF THE RISK OF
BIRTH DEFECTS. (2) CONSUMPTION OF ALCOHOLIC BEVERAGES IMPAIRS YOUR ABILITY
TO DRIVE A CAR OR OPERATE MACHINERY, AND MAY CAUSE HEALTH PROBLEMS.
NO SULFITES ADDED. MAY CONTAIN NATURALLY OCCURRING SULFITES

DOMAINE
Pinnacle.
Cidre de Glace Mousseux / Sparkling Ice Cider
375 ml
RÉCOLTE
2012
PRODUIT DU QUÉBEC   12% alc./vol.
PRODUCT OF QUÉBEC

2014
entre
PIERRE &
TERRE

POIRÉ
DE GLACE
Ice Perry
Vin de poire | Pear Wine
Produit du Québec
Product of Québec
12.4% alc./vol. 1200 ml

TASTING DESSERT CIDERS

Sweet, tart, fragrant, and highly alcoholic, dessert ciders include ice cider, late-harvest cider, and pommeau — a mixture of apple brandy and cider.

1. HEIRLOOM BLEND

Eden Specialty Ciders, Newport, VT

The first commercial ice cider maker in the United States, Eden is an orchard-based producer with its eye on the frozen prize, like this blend of traditional and heirloom cider fruit for the bottled elixir. Passion fruit, orange, and baked apple aromas finish full with a bright acid lift. 10% ABV

2. POMMEAU

South Hill Cider, Ithaca, NY

Sourcing from abandoned orchards along with heirloom apple growers in New York, cidermaker and apple forager Steve Selin takes fresh, unfermented apple juice for this bevvie and blends it with apple brandy. The outcome is a cornucopia of warm, spiced apple flavors, bright enough for an aperitif and mighty fine for a digestif. 19% ABV

3. POMONA

Sea Cider Farm and Ciderhouse, Saanichton, BC

This sipper from the orchard-based cidery was inspired by the Roman goddess of apples. The concentrated juice of this single-varietal crabapple is frozen, then fermented into a high-octane, stone fruit–forward sticky that is both tart and lush. Cheese, please. 17% ABV

4. POMMEAU

2 Towns Ciderhouse, Corvallis, OR

This apricot- and hazelnut-accented tipple is built on acidic and tannic bittersweet apples, fermented, fortified, then aged. The full-bodied drink is finished in a variety of wine barrels, locking in the flavor and bringing in many dimensions that pair beautifully with crème brûlée. 19% ABV

5. NEIGE PREMIÈRE

Domaine Neige, Hemmingford, QC

Fragrant, viscous, and refined, this is the original ice cider from the world's first ice cider maker. Although the cider has adapted over the years, its consistency in quality has not faltered — a golden syrup of tropical fruits, rich apple, and candied citrus cleans up with a sharpness that longs for fruit pie. 12.5% ABV

6. POMMEAU

Westcott Bay Cider, Friday Harbor, WA

In a collaboration with its on-site sister distillery, this aperitif is produced by blending San Juan Island Distillery's apple eau-de-vie with fresh apple juice from the joint orchard and is aged in white oak barrels to finish. Decadent and robust, this cider is divine with Camembert-style cheeses. 16% ABV

7. POMMEAU

Big B's Hard Cider, Hotchkiss, CO

This dessert sipper starts by fermenting and aging estate Winesaps, then that fermented juice is distilled into a 140-proof eau-de-vie and is blended back into cider and apple juice, then aged for an additional 12 months in bourbon barrels. A lengthy process made worth it, the result is a smooth cider of butterscotch, ripe pie apple, and baking spices. 14% ABV

8. TIETON FROST

Tieton Cider Works, Yakima, WA

In the land of many apples, this Yakima Valley cider house picks Jonagold, Pinova, and Winter Banana apples late in the season for this late-harvest cider. Unctuous and rich in ripe, plump, stone-fruit flavor, pair with fresh chévre and honey. 11% ABV

9. SPARKLING ICE CIDER

Domaine Pinnacle, Frelighsburg, QC

Orchard, maple grove, distillery, and cidery, this picturesque producer is placed below Pinnacle Mountain with some tree roots dating back more than 100 years. As the world's first sparkling ice cider, this is an elegant and delightful sipper full of ripe fruit that is balanced with sweetness and dazzling acid. 12% ABV

10. POIRÉ DE GLACE

Cidrerie Entre Pierre et Terre, Franklin, QC

This husband-and-wife team has been making cider in Québec's Montérégie region since 2003 and is one of the few makers on the planet to make ice perry. Using the cryo-concentration method but with late-season pears instead of apples, this tipple is luscious and mouthwatering with spicy pear and tropical fruits. 11% ABV

> Cider on beer, never fear; beer
> upon cider, makes a bad rider.
>
> — English Proverb

HOPPED CIDERS

Humulus lupulus is a familiar plant for both beer lovers and cannabis aficionados. Also known as the hop plant or hops — a key ingredient in beer production with a cult following and a cousin to hemp and marijuana — the fragrant perennial has also made its way into cider.

THE HISTORY

Hops arrived in New England around 1630, presumably carried by the Puritans to make beer in the New World, but it took almost 400 years to make its way from beer to cider. Since the first commercial hopped cider came into production from Salem, Oregon's Anthem Cider in 2010, scores of producers have followed suit, turning a potential one-off trend into a prosperous phenomenon.

The essential oils and bittering acids found in the hop flower (usually referred to as a "cone" or "strobile") provide the signature aromatics that are so sought after for brewing and cidermaking. Originally, hops were introduced into beer production to prevent brews from spoiling so that they could be stored for longer periods of time. Today, it's all about the flavor — North American growers from central New York (the nineteenth-century center of hops production) to the inland Pacific Northwest (which now grows almost 90 percent of the hops in the United States) cultivate 44 different types of hops, each with its own distinct and defining characteristics.

THE BREAKDOWN

Thanks to the generations-long booming hop trade in the Pacific Northwest, cideries from this area were the first to jump on the bandwagon as ingredients were coming from their own backyard. Even with the majority of the hops coming from the same origin, different versions of hopped cider are seen across North America — from slight to puckering succulence.

The spectrum of hop characteristics ebbs and flows through a diversity of aromatic and flavor profiles, like tropical fruits and fresh-squeezed lemon to pine and lavender. Beauty is in the eye of the beholder, with each variety of hops offering different outcomes and adaptations to an apple-heavy profile. Dry-hopping is the main route to take; with this method, dry cones, pellets, and even oils are added to the cider after its initial fermentation. Less common is wet-hopping, in which the "wet," or fresh, hops are added during the primary fermentation. Fresh hops are simply harder to come by because of the commercial demand for shelf-stable dry hops and the need to use them within 48 hours of harvesting.

Although some traditional producers might steer clear of the use of hops in their ciders, the plant can provide a familiar aromatic indicator in an oftentimes less aromatic profile of cider. The use of hops can also complement existing characteristics of cider, like citrus qualities and acid. As an end result, when used in balance, hops can elevate cider to another level.

Hop Sensations

James Kohn, Wandering Aengus Ciderworks and Anthem Cider, Salem, Oregon

"What is the right kind of cider? That's really dangerous to say right now, because there are so many new ciders coming up. If you define for them 'This is cider,' then that's going to hamper them."

Outspoken and meticulous, James Kohn's advice on defining both cider and perry resonates within the industry at such an impressionable time. "Those local producers can really create the space of cider in their markets," he continues, referring to cideries and their local communities. "If there's some authoritative body coming down saying 'This is this kind of cider,' it limits that creativity."

Kohn is one of the founding board members of the United States Association of Cider Makers, the industry-based national organization for cider and perry, and the creator of the annual Cider Conference, the largest trade conference for cider and perry in the country. He faces a lot of opinions, judgments, and perspectives from the industry and consumer on a daily basis. As the co-owner of Wandering Aengus Ciderworks and Anthem Cider in Salem, Oregon, he also asks the same contentious question of his ciders.

Initially, he thought there should be styles of cider, "but it's so controversial," he laments. "There's kind of two measures of flavor profile: there's the intensity, which

cidermakers like, but there's the hedonistic scale of how much consumers like that style. . . . Styles should be based on consumer preferences, not makers' intensity."

For Kohn, Oregon-grown and -made ciders vary in profiles and intensity, from the more traditional-leaning Wandering Aengus with heirloom fruit to Anthem, which exclusively uses dessert fruit and experiments with added fruit and adjuncts, like hops.

Defining those ciders started in 2005 with Wandering Aengus's orchard-and-fruit-driven production, and Anthem launching five years later from a challenge to make solid cider from dessert fruit; employing the same wine-influenced production style the team took with Wandering Aengus. And they succeeded — particularly with those aforementioned hops.

Taking a suggestion from a trusted beer distributor representative, Kohn and his team realized they could run with their proprietary fermentation philosophy but lace dessert apples with hops for a completely novel affect, creating the first-known hopped cider.

Located in the Northwest's epicenter of craft beer — with a perpetual thirst for hyperhopped pints — the addition of hops to Anthem's amiable profile was a no-brainer.

After experimenting with different boils of hops and different varieties, Anthem set its style on Cascade dry hops for the inaugural hopped cider. "After that, everyone else followed suit. In the Northwest, it's a standard to do a hopped cider now. . . I don't believe it's a gateway but it filled a good niche, it's a curiosity to the beer drinker."

Categories, styles, and definitions aside, Kohn believes these variations speak to the diversity of not only the consumer but the drink itself. "To an effect, any of these kinds of ciders are legitimate if someone out there is enjoying it," he says. "It has to be diverse and legitimate. Once it is, then we'll have a sign at retailers that says, 'Beer, wine, *and* cider.'"

"

It's such a simple idea to put hops in the cider, but it didn't really get there until someone outside the cider world suggested it."

Progressive Traditionalism

Tim Edmond, Potter's Craft Cider,
Free Union, Virginia

A constant debate in the cider industry is defining what "real" cider is. Heated and controversial, with a horde of opinions voiced with conviction, the elucidation of "real" cider might not ever be clear. For Tim Edmond of Potter's Craft Cider in Free Union, Virginia, he believes the conversation of cider should be regarding the labels of "craft" over "real."

"For us, we've been trying to give an answer to the question of what 'craft' cider is," says Edmond, to the point that he and his business partner, Dan Potter, put the word in their business name to explain why their cider costs what it does. "It's because of the apples — we get all of our fermentable sugars from the apples. There's an educational component that has been happening. And as the category gets more robust, there's more room for craft. That's where there is durability."

It's always been about craft for the two Princeton University graduates, who forged their friendship in college over disc golf, Neil Young, and craft beer. After brewing in dorm rooms together led to brewing on the weekends once the real world hit, Potter moved to Virginia and the two attempted a farmhouse brewery, with barley and hops on-site, and Edmond commuting from Washington, DC.

It was an experimental fermentation of nearby fruit in 2009 that turned their heads toward something they had disregarded before. Once hooked on Potter's dry cider of Virginia apples, the two were determined to work with the old apple varieties unique to the hills outside of Charlottesville with a traditionalist approach but modern craft slant.

"We've really benefited from there not being a really cogent playbook in terms of how to make cider," Edmond says. "It felt like the beginning of the craft beer movement when there was a lot of big players and not a lot of creative energy being thrown at it."

Taking the blank slate of cider innovation and adding what they knew, one of Potter's Craft Cider's first diversions away from the traditional formula of apples-plus-yeast was with hops.

"We have an affinity for hops, and we've found certain apples work well with certain hops," Edmond says, noting the cidery's quarterly single-hop releases. "We are super juiced about a Sorachi Ace hop cider. It's a Japanese hop with this lemongrass-dill thing happening that's almost culinary. . . . It works really well with GoldRush apples because of the acids that complement the food flavors."

Grateful for their location in the historical region once dubbed the "Apple Belt," Edmond and Potter focus their energy on the local aspect to hone their craft. If there are ingredients outside of the apple sources, they bring them in from neighbors, friends, and the greater Charlottesville community.

"One of the reasons why we settled here was because of the progressive, forensic 'Where does my food come from?' on a real level," Edmond explains. "There aren't that many jobs; a lot of the young people are farming, making food, and cooking. It's a really good community to source ingredients."

While the debate between "real" and "craft" cider goes on, Edmond says the goal for Potter's is to work with the fruit and allow it to take center stage. "Beside the fact that we are using additives like hops, the real focus is letting those apples and yeast working together and [ensure] nothing will overwhelm that apple flavor," he says.

The apples Potter's uses are unique to Virginia, favorites of politicians past or newer "hybrid" apples that can reach both cider production and the fresh market, like GoldRush. For Edmond and Potter, they are advocating the planting of these apples throughout the Shenandoah Valley and Blue Ridge Mountains range, declaring the old Virginia apples from the past as the future for craft cider in their state.

"

We are super juiced about Sorachi Ace . . . It's a Japanese hop with this lemongrass-dill thing happening that's almost culinary."

Tasting Hopped Ciders

TASTING HOPPED CIDERS

Hops add another level of flavor to hard cider — from piney to citrusy —depending on the variety used.

1. MOSAIC HOP CIDER

Potter's Craft Cider, Free Union, VA
The first release in Potter's single-hop series is centered on Mosaic, an aromatic, tropical fruit–focused variety. A match for the mango and pineapple tastes and zingy acid that comes from the apples used, this cider is hop-forward but warmed with those beachside fruits. 8.5% ABV

2. HOPS

Anthem Cider, Salem, OR
The world's first commercial hopped cider came from arguably the nation's most beer-savvy state and was inspired by blending cider and beer. Dry-hopped with Oregon-grown Cascade hops, the cider's floral hops give off pilsner-like aromas, while the juice creates a friendly, sessionable flavor on the palate. Pair with a greasy cheeseburger. 6.9% ABV

3. INDIA PALE CIDER

Bad Seed Cider, Highland, NY
Six generations and 60 acres of apples make up the history for Bad Seed; add a former brewer running fermentations and you get this "IPC." Fermented to dryness with American ale yeast and Cascade hops, this is a cider take on the outrageously popular West Coast India pale ale style and packs a punch like one, too. 6.9% ABV

4. HALLELUJAH HOPRICOT

Reverend Nat's Hard Cider, Portland, OR
This cidery, based in one of the United States' beer capitols, loves hops so much that it has an annual festival for hopped cider fusions. Taking Oregon-grown Cascade and Amarillo hops, Belgian saison ale yeast, and fresh apricot juice, these Northwest apples get a crisp, bitter, yet fruit-forward hit. 6.7% ABV

5. GRASSHOPP-AH

Colorado Cider Company, Denver, CO
Tenured brewmaster Brad Page found his calling in cider and took a little beer influence with him in this hopped cider that also receives a dose of lemongrass. Citrus and apple aromas and flavors are married in this affable easy drinker. Pop a top with Thai cuisine. 6.5% ABV

6. TIETON DRY HOPPED

Tieton Cider Works, Yakima, WA
From the epicenter of America's hop-growing country comes this versatile cider from the cidery's estate orchards within the lauded Yakima Valley. A blend of traditional and exotic hops go into the cider for a juicy, citrusy, and dry sipper. 6.9% ABV

7. HOP'RAGEOUS

Portland Cider Company,
Oregon City, OR

A British expat and an Oregon native marry, and a cidery is born, with hopped cider to quickly follow. Playing on the aromas and flavors of the Citra hops used, orange peel is added to further augment the citrus characteristic and bright bitterness. 6.8% ABV

8. SMOOTH HOPERATOR

Bull City Ciderworks, Durham, NC

A core item for the maker, this cider takes not one but three different hops (Cascade, Citra, and Galaxy) to fuse aromas and flavors of citrus, flowers, and passion fruit together into the fresh-pressed apple base. Crisp and sessionable, this is a ballpark cider; feel free to pair with a hotdog. 5.3% ABV

9. CATAWAMPUS

Blake's Hard Cider Co., Armada, MI

Dry-hopping to bring together the worlds of cider and beer, this sipper comes from a third-generation farm and cider mill that spans more than 500 acres and stretches to three different locations. Citrus peel and juicy apple match the zested lemon of the hops for a refreshing beer alternative. 6.5% ABV

10. EVERYBODY POGO

Cider Riot!, Portland, OR

Ciders and hops from Cascadia: this urban cidery's thirst-quenching hopped cider uses Hood River, Oregon, apples and Willamette Valley hops. With bubbles easy enough to take down a couple pints, this fruit-forward, floral, and citrus-juiced cider can take on both a football game and Indian curry. 6.5% ABV

ROSÉ CIDERS

A set of seemingly normal apple varieties conceal a separate identity underneath the rosy-hued skin — red flesh. An unexpected result for an apple, these varieties are actually more transparent than others, with their insides matching the out. Red flowers to the trees and red skin are flagrant markers for a majority of the red-fleshed apple varieties that have survived extinction on this planet, fighting for their lives in a culinary world confused with their existence. Growing feral worldwide, red-fleshed apples have been cultivated in Europe and Asia for centuries, while in North America, it was a few curious horticulturists that highlighted the apples in the late 1800s.

Rosé ciders are often made by blending a red-fleshed variety in the mix with other apples, or by adding in coloring by another method (macerating the red skins to get the color, aging in red wine barrels, or supplementing with an artificial coloring agent).

THE HISTORY

Like most apples, red-fleshed varieties are found in China and throughout the Middle East, most coming from the *Malus niedzwetskyana*, or Niedzwetzky's apple, lineage. Rare and endangered, this crabapple tree's count in its Eastern source is low, but the tree did spread to Western Europe by the late 1800s and was exported to the United States a few years later. Plantings began to crop up in North America in the late twentieth century as a couple of plant scientists began cross-breeding — horticulturist and botanist Niels Ebbesen Hansen hybridized the Niedzwetskyana for a darker, more brooding red flesh at South Dakota State College, and Northern California horticulturist Albert Etter experimented with the Surprise, a descendent apple of Niedzwetskyana, which exhibited light skin and a more rose-colored flesh. The Agricultural Experiment Station in Geneva, New York, also received credit for the cultivation and growth of red-fleshed apples.

Today, Geneva, a seedling of mother Niedzwetskyana, is the most commonly used red-fleshed apple for ciders, particularly in Québec and New York. Manchurian or Siberian crabapple is a frequent colorist of rosé ciders due to its garnet skin, along with Dolgo crabapple, both apples deriving from Niedzwetskyana heritage.

THE BREAKDOWN

Establishing with the epithet of rosé cider, red-fleshed ciders are red because that is the way the apples are bred. Chemicals in the apple skin, called anthocyanins, are the true source of the hue. These same pigments are to credit for the color of strawberries, currants, and other berries and are thought to be antioxidants, varying in color intensity with each variety and exhibiting different red tones in each.

The oddity of red-fleshed apples, and apples in general, is that the flavor combinations are startling — certain red-fleshed varieties can reveal flavors of raspberry, watermelon, and even buttered popcorn.

As so eloquently put by Tim Larsen of Snowdrift Cider, he realized "that foods are just very dynamic things. They have a lot of compounds that are common across all foods we eat, and depending on the arrangement of those compounds in a given food, you get certain flavor profiles."

The late Terry Maloney of West County Cider was the first to produce a red-fleshed cider in the United States, fermenting the deeply crimson Redfield apple that he found while perusing the selection at the Cornell Agricultural Experiment Station. Redfield served as inspiration for many rose-colored ciders, and Terry's son and current West County cidermaker, Field, carries on the tradition of bottling this rare, ruby variety.

"The apple is our democratic fruit."
—James Fitz, pomologist

Proprietor of Pink

Michel Jodoin, Cidrerie Michel Jodoin, Rougemont, Québec

Nearly two decades before teetotalers began pushing Prohibition across the continent, the Jodoin family had been making cider in Québec's Montérégie region. Beginning with an orchard planted by patriarch Jean-Baptiste Jodoin, cider was produced by the family from the apples grown on the sandy soils of the province's lead agricultural territory. Today, Michel Jodoin, fourth-generation farmer of his family's land and maker behind his namesake cidery that helped put red-fleshed apples on the map, continues the tradition that was once forbidden.

After a vote by the citizens of Québec in 1919, cider — along with beer and "light" wines — was excluded from a list of banned alcoholic beverages during Canada's brief temperance act. A year later, the prohibition was repealed, but a legislative oversight did not lift the ban on cider, which went unnoticed until 1970. After being stuck in a void for 50 years, the situation was corrected, and this great apple-growing region experienced a massive influx of subpar cider production that tragically flopped. Small growers were trying to make good cider on an industrial level; the result was either chugged by a younger generation or dumped down the sink by the elder — few producers survived.

Rising from the ashes, makers like Jodoin have been working against this misconception ever since. "Because of that, we wanted to do something better than people expected," he says, acknowledging that he was one of the first to influence the nuance of cider in the country. "After that, every cider house in Québec had to be as good."

With a rich history of orcharding and cidermaking, Jodoin began making the Michel Jodoin ciders in 1988, pulling from the influences of other cider-rich locales he admired like Normandy and the United Kingdom. Clocking in with three decades of production, the long-term orchardist modestly admits he has had "some success" along the way. The ciders are revered by local wine critics and complemented by chefs throughout some of Québec's best kitchens, and the orchard cider house proves to be a year-round draw not only for Canadians but to international tourists as well.

"The emphasis on the tourist aspect helped to repair the reputation of cider," the French-speaking Jodoin says. "You have to invite people out to taste and explain to them how it's made. It's not just apple juice. When it's made correctly, it can taste great."

Outside of his traditional Champagne-method sparkling ciders and his vibrantly fresh still ciders, Michel Jodoin's most celebrated claim to fame are his red-fleshed cider offerings: six, to be exact, towering over anyone else in North America. Based around the Geneva crabapple — which sports a pink shade any six-year-old girl would love to wear and a powerful tang any cidermaker would drool over — Jodoin fell in love with the acidic, red fruit of the

apple and planted it in a third of his 60-acre orchard, which he believes to be the largest planting of the variety in North America.

Jodoin says he saw past the beauty of the red flowers and cherry-colored hue, straight into the tart fruit and good tannin that a cider would flourish in. "At first, I thought red-fleshed apples would be fun, but I was interested in the fact that you could do a rosé with the apple, without adding cranberry, strawberry, or raspberry juices," says Jodoin. "I stuck with the characteristic of the color; it gave cranberry and raspberry but naturally. . . . It was rosé cider, but natural."

In an area where agritourism is the key to success, Jodoin's cider house sees an average of 43,000 visitors annually. "We have a lot of European visitors looking for something like wine," Jodoin says. "We always have the kind of comparison with European wines, so we have to do our best every time. For the rest, it's the nature of it — we have good terroir for the apple."

Jodoin fell in love with the acidic, red fruit of the Geneva crabapple and planted it in a third of his 60-acre orchard, which he believes to be the largest planting of the variety in North America.

Apples with Intention

Tim Larsen, Snowdrift Cider Co., East Wenatchee, Washington

Maker of one of the most widely distributed red-fleshed ciders, Tim Larsen admits his first attempt at cider in general was an "abysmal failure." The co-founder and cidermaker of Snowdrift Cider in East Wenatchee, Washington, married into a multigenerational farming family, working a 60-acre property that has spanned fruit varieties from apples to pears and cherries since the 1940s. In the 1980s, big business hit this fruit-tree-saturated community, changing the scope of the industry drastically and forcing growers to come up with other purposes for their fruit.

"To be the person that owns the lands and works the land doesn't really work in today's economic environment," Larsen explains. "We loved the land. We asked what we can do with our small farm and ultimately settled on looking into hard cider."

After meeting defeat in the first trials of making cider by using the dessert apples they had on the orchard, Larsen, his wife, and his in-laws got lucky when they found a winemaker who had planted traditional cider apple varieties and decided he didn't want to make cider. "We got the apples, gave it a shot, and we were blown away. It was night and day for us of what we could accomplish if we had the right apple varieties," Larsen says. "One of the most important things to us is sticking to varieties that don't have to be traditional cider varieties but are varieties used with intent and purpose, not just using whatever apples that happen to be around."

Snowdrift's ciders are fueled by the same aspirations today. There are 35 different varieties of cider apples, including red-fleshed fruit, on the property — an epic feat from the beginning. Although the farm has grown and the culture has changed, Larsen and his in-laws, the Ringsruds, carry on the boutique family farm mentality — but with cider instead, what he describes as "farm-to-table cider."

"It's not enough to be a farm-to-table cidery. We have to offer a qualitatively different experience in the cider, and I know we do with the cider apples," Larsen says. "Since the day we've started, we've known that education is the number one thing for growth in this business. . . . That piece is really important to talk about the different types of products, not in a way that disparages them but educates what the difference is in a cider that comes from a farm and one that comes from leftover fruit from the fresh eating industry."

One of the farm-to-table ciders Larsen bottles is the Red Cider, a crimson-colored beauty based on their vast selection of red-fleshed apples that are only "a couple generations removed from their native ancestors" of Kazakhstan, the apple's presumed birthplace. When Larsen discovered these varieties, they rocked his perception of the apple in general.

"I was first drawn to the red color, which is very unique and stunning, but when you start playing with them, you realize this apple tastes like watermelon, this one like raspberries, and this one like strawberries," Larsen observes of the varying flavor profiles of the red-flesh apples. "The fact these apples taste like a watermelon or a cranberry is because they are very similar to cranberries and watermelons when you get down into them. That really drove home to me the value of finding the right fruit or the right apple for the cider."

Simply by choosing the right varieties for the profile he wanted to achieve, Larsen was able to create diversity in his ciders — a notion that he finds authentic to the farm-to-table mantra. "Tannin is not the only tool in a cidermaker's playbook; there are a lot of other polyphenols that are really important in making a good, flavorful cider," Larsen says. "I think that's part of where these red-fleshed apples are really eye-opening. Knowing the varieties you work with is important. Blending brings out the different nuances and bringing better balance to the overall cider."

"
Tannin is not the only tool in a cidermaker's playbook; there are a lot of other polyphenols that are really important in making a good, flavorful cider."

Tasting Rosé Ciders

TASTING ROSÉ CIDERS

In addition to creating a cider with gorgeous color, red-fleshed apples also contribute unique flavors, from raspberry to watermelon.

1. RED CIDER

Snowdrift Cider Co.,
East Wenatchee, WA

Grower-producer Snowdrift has been working with a variety of red-fleshed apples for years. This blend is a showcase of those, both in color and the apple's versatility for flavor, like juicy strawberry, watermelon, and cranberry, slapped with brilliant acid. 7.5% ABV

2. ROSÉ SPARKLING

Cidrerie Michel Jodoin, Rougemont, QC

A signature style for the Montérégie cider house, the rosy-tinted Geneva apple is fermented in the traditional Champagne method for sparkling wine and aged on the lees for 15 months. Fresh apple, peach, and red berry aromas transfer to the palate, which is medium bodied but light and bright. 7% ABV

3. REDFIELD

West County Cider, Colrain, MA

From a house known for its single-varietal ciders, this focus does not deviate with the Redfield bottling, a deeply red-skinned and -fleshed hybrid apple native to the United States. The first commercial rosé cider, its dense red fruit is lifted up with acid and moderate tannins. 6.7% ABV

4. GLOW ROSÉ

Alpenfire Cider, Port Townsend, WA

Pretty and pink from the Airlie Red Flesh apple variety (also known as Hidden Rose), the offering from this peninsula-based cidery is as advertised — glowing. Radiant in color and on the palate, off-dry tropical fruit flavors are zested with acid and orchard freshness. 6.9% ABV

5. CIDER ROSÉ

Uncle John's Hard Cider, St. Johns, MI

Niedzwetzkyana, Geneva crabapples, and a number of other red-fleshed hybrid varieties are blended together in this fruit-forward yet gently spiced and floral sipper from the grower-producer. Try the limited bottling with spicy charcuterie. 6.7% ABV

6. CID ROSÉ

Les Vergers de la Colline,
Ste-Cécile-de-Milton, QC

Produced from macerating the dark red skins of the Dolgo crabapple, this cider might not be red fleshed, but the time on the tannic red skins produces a bright blush sipper. Refreshing strawberry and watermelon reveal an off-dry, palatable pink option. 6.5% ABV

7. THE MAD RUSSIAN

Eastman's Forgotten Ciders,
Wheeler, MI

Some 40-plus red-fleshed apple varieties, like Giant Russian crab, Scarlett Russian, and Alamata, are grown on this estate and pressed into this dry, garnet-colored cider. Three generations of growing old, forgotten apples brought upon this blend of tart berry flavors and gleaming acid. 6.9% ABV

8. BULLE ROSÉ

Domaine Neige, Hemmingford, QC

The ice cider king moves from the sweet to the strawberry with this Geneva-led blend of estate apples. Brunch in a bottle, this *méthode traditionnelle* sparkling cider is fresh in red summer fruits and effortless acid. 6.5% ABV

9. REVELATION MOUNTAIN ROSE

Reverend Nat's Hard Cider, Portland, OR

Another reason to throw a cider party, this cidery hosts a "red fleshtival" celebrating ciders made with the specific fruit and makes this bright and floral small-batch sipper with the rare, namesake variety. Pair it with a tangy-sweet summer salad. 6.4% ABV

10. CRÉMANT DE POMME ROSÉ

Cidrerie du Minot, Hemmingford, QC

The small, red-fleshed Geneva apples are featured in this tangy cranberry and tart apple cider. The tipple's natural effervescence calls for fried chicken and waffles, and its low, low alcohol content calls for another bottle. 2.5% ABV

FRUIT-INFUSED CIDERS

Fruit-infused or fruit ciders seem to be the common ground for the varying schools of cidermakers. For some traditionalists, although adding fruit flavor or juice does take away from the apple, it can be done so with other fruits that complement both. For experimental producers, the apple doesn't fall too far from other fruit trees, and fruit-infused ciders offer diversity in an exceedingly thirsty market.

While the debate over what is "real" cider continues, this sidestep doesn't fit into the derogatory, bastardized "apple-pop" or "alco-pop" category if it keeps to the basics of fruit. Fruit-infused ciders start with a fermented apple base and add fruit or fruit juices, either before or after fermentation. By this loose, cross-cultural, and dated legal definition, fruit ciders still abide.

THE HISTORY

It could have been divine intervention from the harvest-proliferating goddess Pomona, a farmer's good sense to utilize all of the fruit he grows, or sheer coincidence, but it is a recent occurrence for fruit other than apples to make its way into a standard cider blend. Supplemental flavors and structural acidity strike a chord with complexity when other fruits are used in dessert apple ferments, piquing interest for drinkers foraging for more flavor.

For fifth-generation family grower Mike Beck of Uncle John's Cider Mill and Uncle John's Hard Cider, he has been growing a variety of fruit on his Michigan orchard since his ancestors came to the site. Like many stories of how cider production began on an orchard, his fruit-infused ciders came from employing the crop at hand.

In the early 2000s, Uncle John's worked closely with Michigan State University on one of the school's initial studies into cider research, collaborating on the production of a cherry apple cider — one of the first to surface on the market and a sought-after product from the cidery today. In recent years, modern and experimental cidermakers are bottling, canning, and putting these ciders on draft, and the consumer is sucking it down.

THE BREAKDOWN

Big apple-growing regions in North America are big fruit-growing regions across the board: Washington State grows a lot of cherries and raspberries, Oregon stakes the claim for pears, Michigan's blueberry and cherry crop is huge, California boasts plums and strawberries, Virginia is the seventh-largest supplier of table grapes, New York flourishes in stone fruit like peaches, and the list goes on.

Pear ciders — not to be confused with perry, exclusively made from fermented juice of pears — fall into this category, along with the popular cherry and cranberry ciders, which all take the apple juice base, then add the variable juice or flavor.

As the hyperlocal food movement pushes more into cider, along with hand-crafted products, demand for truly regional tipples will grow louder. For producers who aren't able to get ample supply of apples suited for cider, they might be able to source other local fruits and adjuncts like spices or botanicals, satisfying that need for local in the bottle to their customer.

From cherries, blueberries, cranberries, apricots, prickly pear, passion fruit, and even watermelon, the fruit combinations are endless. For experimental cidermakers, these New World ciders not only add different elements of acid, intensity, and flavor but they boost the growing anthology of the category.

Flushed in Fruit

Mike Beck, Uncle John's Hard Cider and Uncle John's Cider Mill, St. Johns, Michigan

"We were early transplants: the very first of my family bought homesteaded land where we are now," says Mike Beck, the founder of Uncle John's Hard Cider and fifth-generation Michigan fruit farmer. "We had a knack for fruit growing and found a piece of property that was good to do it. That's how we came about being here."

For as simple as Beck makes it sound, his family is one of long-standing Great Lakes fruit heritage, but it wasn't until his parents' go at the farm that Uncle John's Cider Mill was created and converted from an old barn on the property. As a response to a depressed wholesale fruit market, fresh juice and cider donuts were the original business plan, with everything from pies and preserves to honeys and syrups following suit and eventually leading into the creation of fruit-infused cider with their son's generation.

According to Beck, as long as there were fruit trees in the ground of the St. Johns, Michigan, orchard and mill, cider was being made for friends and family. Although Prohibition nixed the potential of selling it to the commercial market earlier in the twentieth century, Beck and his wife, Dede, resurrected the practice in 2002 as Uncle John's Hard Cider. Today, the total acreage of the farm is at 300, with 80 acres in apple trees, and that 120-year-old barn houses the cider production.

"Most of the first ciders were made from fairly common apples, and we had some very nice heirlooms," Beck says. "We have since replanted both American and European (cider) varieties. . . . As a grower, we know so many people in the business, and there's a lot of unique heirlooms in Michigan."

Many people know Beck as well. A farmer at heart but a businessman and spokesperson by trade, Beck was one of the founding members of the Michigan Cider Makers' Guild and of the United States Association of Cider Makers. He is a well-known proponent of cider, of orcharding, and also of innovation.

Beck strives to take advantage of Michigan's vast crop diversity, from his premium line of ciders that lean more toward an orchard-based, European style to his modern, fruit-infused ciders. He credits this range of fruit growing to Michigan's deep processing culture and infrastructure, noting that much of the fruit found in commercial pies or pastries hails from the Great Lakes State.

"We grow a lot of fruits, and we have a lot of things to work with here," Beck says, noting a collaboration with Michigan State University in producing a research-based

cherry-infused cider. "We were one of the first batches they did with that. We are just trying to show another Michigan commodity in a beverage."

Cans of Uncle John's Hard Cider reveal their crop diversity — from apple cherry and apple pear to blueberry, apricot, cranberry, and other seasonal releases.

"Consumers always want to push the envelope on flavors, and my fruit ciders have been around for a while," Beck laughs, suggesting that his are probably considered mundane now and he needs to add something like passion fruit to keep up.

Although his fruit-infused ciders might have a wider variety and larger total production than many on the market, Beck hopes that drinkers will graduate to his premium line of ciders, which include a perry and several single-varietal, Michigan-centric ciders.

In the end, other than the infusion of fruit or other adjuncts, Beck says, it's all about the apple. "That's paramount to me."

Cans of Uncle John's Hard Cider reveal their crop diversity — from apple cherry and apple pear to blueberry, apricot, cranberry, and other seasonal releases.

Community Support

Trudy Davis, Eaglemount Wine and Cider,
Port Townsend, Washington

Crowdsourcing isn't just for fund-raising and finding cat photography online. It's also how Trudy Davis makes each of her ciders, and has been doing so for more than 20 years.

"I feel like we're pretty lucky to do this," Davis says of her Port Townsend, Washington, winery, Eaglemount Wine and Cider. "People bring apples by and we'll trade for fresh cider or buy them. We've gotten to be known locally for doing that."

Established on an old homestead with its original barn raised in 1883, Eaglemount thrives on old, funky apples and other fruit, supplied from backyards and run-down farms, from ancient, forgotten trees throughout the lush and picturesque Olympic Peninsula. After renovating the barn into livable conditions, Davis and her husband, Jim, started wine and cider production small, in an even smaller room.

That one little room was roughly 14 by 10 feet, "and we did about 1,000 gallons of cider and 1,000 gallons of red wine in it," Davis estimates. "A year later, at the farmers' market, we started selling. . . . It was like we were introducing hard cider to this whole area. And we've been doing that ever since."

When cider production launched in 2006, the Davises were getting a majority of their apples and fruit from their orchard

on the homestead and a neighboring orchard, occasionally having to dip into eastern Washington for more. However, in the most recent years, the Olympic Peninsula has been prolific enough to be the exclusive supply of unique, old apples for Eaglemount. Davis says she employs a "hunter-gatherer" to find these unused fruit sources and then trades with cider or juice to help pay for the apples.

"All of our ciders are made with these old apples from the community," she says. "When you get juice from these old trees, it's so good, you just know it's going to make good cider, too."

Some of these heritage and heirloom apples (including Gravenstein, Winesap, White Pippin, Winter Banana, and Roxbury Russet) have either disappeared from the grocery stores or never made it into their warehouses in the first place. Their dormancy has been Davis's key to success — using community-sourced fruit that was going to waste. "It seemed like the right thing to do with the apples," Davis says.

Good to the land and good for cider innovation, Eaglemount was the first commercial cidermaker on record to make a ginger cider. Davis hasn't stopped there, with other infusions like a quince cider from estate trees, a rhubarb cider from plantings on-site, plus a raspberry ginger and raspberry-hopped cider and several fruit-infused meads.

"I love the traditional ciders, but there are so many things you can do with cider, like the ginger cider, or hops, or different fruits," Davis notes. "It's just a whole new world."

Davis thinks her fruit-infused ciders and ginger-spiced sippers increase the diversity of cider. "I think these ciders fill a niche, adding variety," Davis says. "People have so many different taste profiles that they like. There's always the favorite of the traditional, but adding something in is just the frosting on the cake. . . . Specialty ciders are just an addition to the whole spectrum of cider."

Eaglemount thrives on old, funky apples and other fruit, supplied from backyards and rundown farms, from ancient, forgotten trees throughout the lush and picturesque Olympic Peninsula.

Tasting Fruit-Infused Ciders

1 DOC'S DRAFT
Peach
HARD APPLE CIDER
The Great American Cider™
650 ML.

2 CHAMPLAIN ORCHARDS CIDERY
VERMONT HARD CIDER
HONEY PLUM

3 SANTA FE CIDER WORKS
Enchanted Cherry
Artisanal Hard Cider
750 ML. (25.4 FL. OZ.)
ALCOHOL 7.0% BY VOLUME

4 Uncle John's
HARD CIDER
APPLE PEAR
GLUTEN FREE
16 OZ. / 473 ML / 6.5% ABV.

5 finn CID
Black Currant
6.5% alc./vol.
BOTTLED BY HAND
BATCH
Farmcrafted
On the Olympic

TASTING FRUIT-INFUSED CIDERS

From blackberry to pineapple, fruit-infused ciders bring together the best of the fruit worlds.

1. PEACH

Doc's Draft, Warwick, NY

A flagship cidery in the Hudson Valley, this tenured maker has been working the land and turning out fermented fruit (cider, wine, brandies, and liqueurs) since 1993. The juicy and supple peach sipper stays faithful to the apple while still flirting with the honeyed and coquettish flavors of peach in this semi-sweet seasonal release. 6% ABV

2. HONEY PLUM

Champlain Orchards Cidery, Shoreham, VT

A big, aromatic bouquet of honey, meadow, sage, and chamomile meld with plum and apple. Rosy pink, this cider infuses both Vermont honey and estate-grown Methley plums into the Honeycrisp base, sitting clean and juicy on the palate. 5% ABV

3. ENCHANTED CHERRY

Santa Fe Cider Works, Santa Fe, NM

Montmorency cherries in a bottle, sour yet sweet pie-cherry aromas dominate the New Mexico apples pressed in this cider made by this female-owned and -operated cidery. Ultimately, apple prevails on the palate, mingling with the tangy cherry and rhubarb flavors. 7% ABV

4. APPLE PEAR

Uncle John's Hard Cider, St. Johns, MI

A recent year-round addition to the grower's cider portfolio, farm-fresh Michigan Bartlett pear juice is added to the ferment for a delightful fruit-basket flavor that finishes fresh, semisweet, and approachable. A recipient of multiple awards and many good times. 6.5% ABV

5. BLACK CURRANT

Finnriver Farm & Cidery, Chimacum, WA

A blend of heirloom and organic dessert apples, this farm-based cider includes locally foraged organic black currants to complement the light and sprightly cider. Gorgeously purple, the berries bring tartness and tannin to the forefront. 6.5% ABV

6. ROOT RIVER AMERICAN SESSION-STYLE RASPBERRY

Wyndfall Cyder, Jordan, MN

After receiving his master's in horticulture, Rob Fisk founded his cidery and began pressing Minnesota apples on-site for fermentation. Named for the flowing southeastern state river, this cider is juicy in fresh raspberry fruit and refreshingly tangy and tart. Quickly quaffable and very pleasant. 6% ABV

7. CRANBERRY BLEND

Downeast Cider House, East Boston, MA
Proud New Englanders and cider patriots, Downeast cans and distributes one of the few fruit-infused ciders of the region. Fresh-pressed cranberries are added to the fermented apple juice for rejuvenating tart red berry flavor, finishing simply crushable and best when fresh. 5% ABV

8. BRAMBLE BUBBLY

Sea Cider Farm and Ciderhouse, Saanichton, BC
Vancouver Island blackberry lends its hue to this limited-release bottling from the peninsula-based farm and cider house. This cider speaks the truth: brambly blackberry fruit is juicy yet sharp against a warm and pleasing apple back. 9.9% ABV

9. BLUE GOLD

Vander Mill, Spring Lake, MI
Western Michigan blueberry juice is added to this semisweet apple base from grower-producer Vander Mill, showcasing the best of both fruit worlds, for a rich but clean cider that conveys imagery of picking fresh blueberries off the bush. 6.9% ABV

10. PINEAPPLE EXPRESS

Jester & Judge, Stevenson, WA
A seasonal offering for this stunningly located riverside cidery, Northwest apples are melded with tropical pineapple juice, building the succulent body of this unfiltered tipple. Moderately tart and generously plump, the pineapple adds creaminess while the acid zips it all together. 5.5% ABV

BARREL-AGED CIDERS

At its traditional and historical core, cider is barrel aged. That's not a matter of preference or debate — although it might still be controversial, knowing this crowd — it's just history. Cider was originally fermented in oak likely because of the lack of other fermentation vessels — stainless steel wasn't patented until the late 1800s, and concrete was and is largely saved for wineries.

In North America, while in-house barrel aging has been a trend for cocktails and oaked-up Cabernet Sauvignon still reigns supreme, the method is nothing new to the respective homes of Kentucky bourbon and Canadian rye whiskey from which many cideries are collecting their barrels now.

THE HISTORY

The farmhouse cideries of France and England relied on oak barrels not only for aging but also for transportation. Most traditional *sidras* of Spain were barrel-aged and poured from the spout of an oak vessel, typically elevated feet above the ground so that the cider would become effervescent as it flowed into the glass.

When the drink made its way west to New England, it was produced in barrels, oftentimes with feral yeast and whatever else might have been occupying the farm or cellar it was housed in.

Today's cidermakers come to barrel aging through one of two paths: innovation or historical inspiration. Cider is aged in barrels for a score of reasons, from tannin acquisition from the wood and structural formation to oxidation notes and pulling flavors from what was previously aged in the oak, like bourbon, brandy, or red wine. Nowadays, makers are able to skip the barrel and achieve similar effects from oak chips, which cut the price tag to a fraction of what the barrel might cost, depending what was in it before and where it's coming from.

THE BREAKDOWN

A style recognized by the select cider competitions in the United States, "New England" cider is defined by the Beer Judge Certification Program (also known as the BJCP, the nonprofit organization that certifies and ranks beer judges) cider-style guidelines as "a cider made with characteristic New England apples for relatively high acidity, with additives to raise alcohol levels and contribute additional flavor notes." One of the most common additives is the barrel and its impact on the cider itself.

A large proponent of barrel aging, Scott Donovan of BlackBird Cider Works in New York says his New England–style cider was the one that wrote the book for BJCP — it was the first modern commercial example used to explain the style. He uses barrels to "produce more heritage-style ciders," and allows the barrel to give a sense of place as well, be it from Kentucky bourbon, Tennessee whiskey, or a Buffalo, New York, winery.

Cideries are ramping up their barrel-aging programs across the continent as production facilities and resources expand, and as the consumer demands something that straddles cider's disputed identity: a little bit traditional and a little bit innovational.

A Culinary Hand

Chris Haworth, West Avenue Cider Company, Toronto, Ontario

Apples, barrels, Niagara Montmorency cherries, Ontario Merlot, shiso, hops, star anise — the pantry list of potential cider ingredients goes on for former London chef Chris Haworth of West Avenue Cider.

Upon moving from the United Kingdom to Toronto, Ontario, home of his Canadian wife, Haworth's chef roots planted deep into the vast apple-growing region of the province. A home brewer on top of his culinary credentials, the Brit was perplexed as to why cider hadn't taken off with such a healthy crop in Ontario. "The epiphany came: there are so many apples in Ontario, but there's no cider up here," Haworth says. "The rest of the world is making cider, so I started experimenting with the apples that were available, different yeasts, and wrapped myself up in the cider world in 2008. By 2012, we had our first cider on the market."

Although the apple market was flush, the traditional French and English cider fruit was far and few between. Haworth found himself traipsing the countryside in search of the apples he wanted to use in his ciders, 100 percent heritage Ontario fruit like Russets, Baldwins, and Empires. But he wasn't able to get his hands on the bittersweet and bittersharp apples that he was used to seeing in English ciders, which led

the kitchen veteran to get inventive with his ingredients.

"When you don't have apples, you might add a bit of ginger, ferment it with a different yeast. I just searched for flavors, flavors I couldn't get," he says of his first ciders, which included Cider Weisse, a cider comprising citrus, spices, and hops, and a collaboration cider with famed restaurant Momofuku that uses pear, shiso, yuzu, ginger, amazake, and sake yeast.

To remedy the lack of choice heritage fruit, Haworth and his family purchased a 75-acre farm in 2013, planting 52 varieties of apples across 15 acres and testing different varieties to see what works in their soils. "We're experimenting with fruit that's been around for centuries in North America rather than going back to Europe," Haworth details. "I love the European ciders, too, but I'm just trying to find out what's good for me on my soil. . . . I get a buzz from that."

Haworth also gets a buzz from his playful production, which he has toned back a bit since starting the cidery, nowadays focusing on barrel aging, wild fermentation, and growing the right apples for his ciders.

For his extensive barrel-aging program, Haworth sources barrels from Niagara's loaded wine scene, a hot commodity to his production and aging vessel for many of his ciders. "There's so much more flavor that comes out of the barrel," he says. "I kind of let the apples and the wild yeast develop and barrel-age, whatever else bugs and bacteria that are in the barrels, I like that . . . . I don't really monitor the fermentation much, I let it happen. I love how the barrels give the oxidation."

With heritage apple trees in the ground, many of which he plans to press into single-varietal offerings, Haworth likens this focused production to the wine industry. "I'm very much a traditionalist in that sense, but when a traditionalist can't get access to apples, that's when you have to make up for it with ginger and sake," he explains. "You have to get creative, otherwise you get bored."

For the chef, he says, until his orchard is ready to produce enough fruit for the cidery and the single-varietal ciders, he pulls apples from across the province to fill his recipe requirements. Acidity is a major ingredient to his ciders, which he also credits to his culinary background.

"I like that extreme flavor, that bite," he says. "Cider has to be food friendly. That's why I like the acid; my ciders have big acids to cut through the fatty foods, the foie gras and the pork dishes. It holds up well to food, it complements it."

One of his first ciders was a collaboration with famed restaurant Momofuku that used pear, shiso, yuzu, ginger, amazake, and sake yeast.

Homage to Heritage

Scott Donovan, BlackBird Cider Works, Barker, New York

It took Scott Donovan two cross-country treks to find his calling in cider. Raised among agriculture in upstate New York, he moved to the Seattle area for a job with an accounting firm and met cider by way of a fellow home brewer at a party in 2000. Struck by the intensity of the funky, farmhouse aromas and flavors of the cider he was sipping, Donovan soon moved back to his home state and began looking for an orchard to make cider from in Rochester.

"From the beginning, I was really interested in growing apples and making cider," Donovan says of the somewhat downtrodden orchard he found in 2006. "It needed some tender love and care. . . . It was a struggle to learn growing apples, but I truly believe you don't have to be a second-, third-, or fourth-generation farmer to grow apples. We've been successful. You just have to learn how to grow and get your hands dirty."

For the next five dirty and laborious years, Donovan harvested, pressed, fermented, tasted, and learned more about his trial ciders, launching BlackBird Cider Works exclusively from Donovan Orchards to the public in 2011.

The 37.5-acre property has roughly 20 acres of orchards planted, growing 25 different varieties like New York favorites Empire and Jonagold, as well as English cider apples like Kingston Black and Dabinett, some varieties grown organically, pears, and more. Like the orchard, BlackBird's range of ciders is broad and extensive — from ciders made with organic apples and pub-style draft exclusives to barrel-aged tipples and a variation Donovan coined "New England–style cider."

A self-proclaimed cider history nerd, Donovan says the influence for BlackBird Cider Works, particularly in his barrel-aged ciders, comes not from the English or the French but from New York's own historical tradition.

"We strive to produce more heritage-style ciders like how they used to make them in this area 75 to 100 years ago," Donovan says. "This apple valley area has always produced cider, but mostly for personal consumption. The farmers used to get their barrels in the city: they'd buy them at farmer trading markets, used from Kentucky bourbon or Tennessee whiskey. They'd bring them back to their press houses, get them filled, and put these vessels in their garage, basements, sheds, or barns."

BlackBird Cider Works even put this historical hypothesis to the test with a focus group of elderly farmers in the area who remembered buying barrels from these markets with their parents and grand-parents. "That really influenced us in the New England and Kentucky barrel-aged ciders we do. It celebrates the connection between the two regions where the barrels used to get picked up," Donovan says.

Outside of the history nod, Donovan simply enjoys the impact of oak on his ciders. "I think the oak brings out a lot of interesting flavors in the fruit, things you wouldn't necessarily highlight if it was just aged in stainless steel."

BlackBird was one of the first to the commercial market to reinvigorate this "New England–style" of cider, a definition and "style" Donovan admits has been a hazy mission to take on. In its simplest terms, he interprets it as fresh-pressed juice fermented in a wooden vessel to dryness, which then goes through secondary fermentation with adjuncts like honey or raisin to enhance flavor and boost alcohol level.

A heady concoction, Donovan's colonial-style, semisweet cider is a blend of nine different varieties from the farm and takes in both oak and honey for fermentation.

"We're seeing that drier ciders like this are more popular than they were a few years ago," Donovan says of the success of his ciders. "It's a natural evolution of the consumers' palate."

Having gone through an evolution himself from a consumer to a grower-producer, Donovan doesn't view a cider apple as a cultivar that necessarily comes from English, French, or even American heritage. He believes it starts with a good balance of sugar, tannins, and acid, a trio he dubs the "fundamentals" to compose his ideal cider, maybe a cider that pays homage to American farming and cider heritage along the way.

> 66
> The oak brings out a lot of interesting flavors in the fruit, things you wouldn't necessarily highlight if it were just aged in stainless steel."

Tasting Barrel-Aged Ciders

TASTING BARREL-AGED CIDERS

Aging cider in barrels lends tannin and flavor from the wood, along with essences left from whatever beverage was previously aged in it — usually wine or whiskey.

1. LE CHÊNE

Stem Ciders, Denver, CO

Pronounced "lu-shen" and translating as "the oak," this cider takes refuge in used Zinfandel wine barrels to age, mature, and soak up the goodness of the grape's influence in oak. The result from these barrel-stained traditional cider apples is one of layers — smoke, vanilla, and red berry fruit tapped with chewy tannins and easy acid. 6.4% ABV

2. STELLAR

Argus Cidery, Austin, TX

The resourceful makers behind this cidery dig deep to find apples from Texas. They go the extra mile in fermenting this single-varietal cider made with Stellar apples grown in the Southern Ozarks and aging in French oak. The details sum up to a full-bodied, fruit-forward, opulent and tannic drinker. Limited release and vintage-specific. 6.7% ABV

3. CHERRY BEACH

West Avenue Cider Company, Toronto, ON

Nearly lambic and completely unique, this sour sipper ferments heirloom apples with Niagara Montmorency cherries and ages the juices together in Pinot Noir barrels for a butter-cream textured, red apple, cherry, and resin-flavored cider. 7.5% ABV

4. DRY ORGANIC OAK AGED

BlackBird Cider Works, Barker, NY

In honor of New England cider of yore, this orchard-based cidery ferments a blend of organic estate apples, ages it in French oak and allows for the smoky, woody true colors to shine through. The still cider is redolent of a delicate, floral white wine; dry, citrusy and slightly spiced from the gentle oak. 6.8% ABV

5. LAPINETTE

Virtue Cider, Fennville, MI

This Normandy-inspired cider ferments Michigan apples with a French saison yeast strain and ages the juice in Cabernet Franc barrels from California that were inoculated with *Brettanomyces* yeast to develop farm-house funk along with the fresh and yeasty aromas. Bright, crisp, and slightly sharp, this is food cider to a T. 6.8% ABV

6. NORTH MEETS SOUTH

Good Intent Cider, Bellefonte, PA

To commemorate the Battle of Gettysburg, this Northern cidery found a host in Southern bourbon barrels for this limited-release sipper. Pennsylvania-grown apples are fermented, then aged in Virginian barrels for a cider that is bone-dry with soft bourbon and vanilla aromas; smooth and fruit-forward flavors mask the higher alcohol. 10% ABV

7. PIRATE'S PLANK "BONE DRY"

Alpenfire Cider, Port Townsend, WA

From the state's first organic cidery, this bone-dry cider was roused by similarly styled English ciders. Estate-grown English bitter-sweets and bittersharps (plus some Granny Smith) are fermented into an unapologetically tannic and sharp cider with aromas and flavors of red apple and earthy aromas of hay, dusty skin, wax, and citrus peel. 6.9% ABV

8. SEMI-DRY

Farnum Hill Ciders, Lebanon, NH

The majority of the cider at this orchard-based producer is aged for a lengthy period in neutral oak barrels in order to absorb the structure and woody tones from the used brandy and bourbon barrels. This offering reveals elaborate aromatics with subtle fruit-forward, oak-accented flavors that are zapped by bittersweet acidity and tannin. 7.4% ABV

9. KINGSTON BLACK

Distillery Lane Ciderworks, Jefferson, MD

The prized possession of Somerset, England, this estate-grown cider apple is fermented to dryness, then aged in bourbon barrels from Virginia. Fresh apple, stone fruit, and honey are kissed with faint oak aromas, while smooth vanilla and ripe apple fill out the palate. 7.5% ABV

10. STONEWALL

Liberty Ciderworks, Spokane, WA

General "Stonewall" Jackson, alleged lover of mixing bourbon and cider, receives a nod with this limited-release tipple from the eastern Washington maker. Borrowing whiskey barrels from a neighboring distillery, Liberty's oak-soaked cider is approachably dry with wood, vanilla, and baked apple flavors, finishing with a nourishing, heady zing. 8.5% ABV

SPICED AND BOTANICAL CIDERS

When baking an apple pie, gourmands shuffle through their pantries in search of accompaniments like cinnamon, nutmeg, clove, and other sweet and savory spices to enhance the flavor profile of the baked apple. In the kitchen, chefs thumb through herbs and botanicals like juniper, lemongrass, and rosehip that complement their dish. Cider, an expansive palate of fruit and flavor, is also an open palette for innovative design, welcoming this type of culinary chemistry to the fermenter.

THE HISTORY

A recent category of cider variations, spiced and botanical ciders join the list of options for modern cidermakers — many hailing from the beer industry, either on the producer or the consumer side — to explore what they can do with an accessible and sometimes simple dessert apple base. Turning to natural agricultural accessories for cider by adding spices or herbs to the ferment or in the final blend is a trend borrowed from beer, and one that is bringing drinkers of suds along with it.

In locations where apples for cider have to be sourced regardless, cideries look to other agrarian means like spices and botanicals to add regional flair, originality, and flavors to diversify not only their cider selection but also the offering from a retail shelf to the consumer. Whimsy and ingenuity combine in these creative takes on the apple and its addition of flavors.

THE BREAKDOWN

Some makers of the variation say they pull inspiration from the culinary scene, using ingredients from a baker's basic pantry, the bins of a farmers' market, or a bundle of herbs making its way to the still for gin. For Crystie Kisler of Chimacum, Washington's Finnriver Farm & Cidery, her prior work experience on an herb farm gave way to fully grasping the plush herbal jungle of the Olympic Peninsula. Today, she resourcefully uses ingredients like lavender, spruce tips, and honey meadow from her own surroundings.

Outside of the apples, makers are curating and creating ciders unique to their palates, regions and pantries, allowing the adjuncts to also generate a sense of wellness if a sense of place isn't as easily attained through the apple market. The health benefits that come from cider alone (rich in disease-fighting antioxidants, high in vitamin C and naturally gluten-free) are elevated with those of spices and botanicals, like digestive-relieving ginger and lemongrass, muscle-relaxing quinine, blood-sugar-lowering cinnamon, skin-enriching rose petal and whatever else the inner botanist desires.

Spices can complement the ripe, juicy fruit of a dessert apple cider base and match the autumnal flavors in foods of the same nature. Botanicals can adhere to the more savory side of the apple, revealing gin-like qualities and matching with harder-to-pair foods like greens.

Sense of Place

Crystie Kisler, Finnriver Farm & Cidery, Chimacum, Washington

Historically, traditionally, and rurally speaking, cider has always been about community and the land. An agricultural outcome of the soil in which it stems from, cider became the microphone that Keith and Crystie Kisler needed to share the story of their place.

Keith Kisler, a descendant from a wide family tree of farmers in eastern Washington, met Crystie, an East Coast transplant, while working in California's Yosemite National Park. The two taught younger generations about nature and saw a separation between what kids were learning in the park and how they practice their lives, with the land and food, at home.

"We saw that food is nature and we needed to start integrating a conversation about food into our discussions of the environment," Crystie Kisler says. "We decided we would become farmers and create a farm that would help reconnect people to the land and restore a healthy environmental ethic."

With no expectations but high standards, the Kislers bought 33 acres of fruit trees and more in Washington State's Chimacum Valley, and Finnriver Farm was born. Once on the land, reality hit, and they realized they needed to fund their dream. After experimenting with CSAs, U-pick farming, and other potential business models for the orchard, they were brought a bottle of home-fermented cider from their neighbor and the plan was settled — they would produce cider from the farm.

"At that point, we were completely unaware of any cider movement or trend," Kisler says of when they started plans for the cidery in 2006. "For us, it seemed like a natural outcome of what was growing on the land."

The farm quickly grew as the team learned they needed more apples than they were growing in the orchard out back. Planting another orchard of roughly 800 trees on the original farm, and leasing a 50-acre property in addition, Finnriver Farm & Cidery now sits at more than 6,000 apple trees and 20 different varieties across the two lots of land.

"We did what felt right and seemed like it would work, being true to our heart and our taste buds," Kisler remembers. "Essentially, cider saved the farm, and now we try to serve the farm with cider."

With the farm as the base to the cidery, Finnriver's commitment is to use the apples — and other farm products such as raspberries, honey, squash, lavender, and more — to create an exceptional product in the bottle and to promote what Kisler calls "a healthy agricultural ecosystem."

This dedication to the land shines through in the scads of ciders Finnriver produces, particularly in the Elijah K. Swan Seasonal Botanical ciders that showcase the cidery's innate ability to work with nature. "We set out to discover what could be fresh and ripe on the land and communicate the season," she explains. "We wanted to actually capture that. It was seeing what was actually growing outside on or near our landscape within that season."

The botanicals line tells the story of a year, 365 days of cider. With a honey-infused cider in the spring, Black Currant Lavender for summer, Forest Ginger in the fall, Cranberry Rosehip for winter, and more, including Saffron cider that sources the spice from a nearby grower, the line is ever evolving. What they aren't able to grow themselves, Finnriver locates from other sustainably minded farmers, and as locally as they can. The Kislers' children, along with fellow farmers and neighbors from within the community, pick the fir tips, lavender, and rosehips for the ciders. This "agricultural rhythm," as Kisler labels it, is a defining characteristic in Finnriver's ethos and overall message.

"We are making something that people put in their mouth. I can hardly think of a more intimate act," Kisler says beaming, confessing she usually does a little dance when she talks about this. "Let's get the most beautifully grown fruit that has been created with the intention to do no harm to the earth or human body, and craft that into the most careful, considerate, and conscientious cider we can make so when you put it in your body, it does some magic. . . . We cherish that intimacy, that sort of privilege and responsibility."

> 66
> We set out to discover what could be fresh and ripe on the land and communicate the season."

Foraged and Found

Sean Kelly, WildCraft Cider Works, Eugene, Oregon

Sean Kelly is a forager by nature. A woodworker of salvaged materials by trade and a student of the ancient herbal-based Ayurvedic medicine, he sees the potential in the wild and in the natural habitat surrounding him in western Oregon. He has found himself seeking out the abandoned and uncultivated forests his area of the state provides, just 50 miles east of the Oregon Coast, south of Pinot Noir wine country in Willamette Valley and at the confluence of two major rivers in Eugene. Surrounded by abundant flora and fauna, Kelly took his passion for using his environment and fermented it.

WildCraft as a concept is the result of "a forager mentality," Kelly says of his cidery, combined with a business plan that formulated after years of making botanical wines. "That's something I've always been attracted to — utilization."

He began by exploring what flowers and plants could offer in fermentation, the potential medicinal properties they could provide, and the bombshell aromatics that would develop out of what seems like nowhere. Kelly was energized by what he could achieve with these fermentations, enhanced by what was available to him in his area. From there, he began to search for old orchards, left to run feral and free for decades, where original Oregon

homesteaders planted varieties fit for cider production and clean drinking.

"Seeing the fruit, the age of the plantings, the history of the different settlements, and the variety of the apples — I really started getting into it," Kelly explains. When approached by a group of supportive investors with the proposition of running a botanical brewery, Kelly countered with a cidery, and they all signed on. "Apple's complexity in its own and how it combines with botanicals was intriguing. . . . The profiles of the herbs add more to the subtleties of a dry cider, with its already delicate nuances."

Kelly's herbal foraging has led him into studies of the yeasts that grow on wildflowers and observations on the range of terroir that can be expressed from different areas of a single field. The organic yeast cultures built up from flowers have inoculated large batches of WildCraft ciders, specifically yeast from apple tree blossoms, which Kelly says generates the most dominant wild yeasts for cider.

WildCraft Cider Works is revered for its small-batch, wild-ferment experimental ciders, where Kelly runs wild through fields of ferments with herbs, flowers, leaves, and more like wild rose, lilac, nettle, licorice fern, and sage. The cider house — which is outfitted with furniture and bar tops built

Kelly's herbal foraging has led him into studies of the yeasts that grow on wildflowers and observations on the range of terroir that can be expressed from different areas of a single field.

by Kelly himself — pours 10 ciders on draft at once, with a new botanical line tapped as soon as the current flavor of the month runs dry.

Although the Alcohol and Tobacco Tax and Trade Bureau puts a few restrictions on bottling and distributing his radical creations outside the cider house due to potential psychoactive components or allergens the herbs might contain, Kelly is able to pour for droves of thirsty Oregonians in-house and educate them with each ounce.

"There is a long-standing U.S. and European history on botanical fermentations," he says, documenting the history of herbal beer and flower wine as an embedded part of farmhouse culture. "It's just an amount of time before they take root. . . . Oregon as a region was planted with cider fruit, but maybe not the kinds that were planted in New England or England, but the kind planted by the settlings here for consistent production for safety, for clean water, to be able to ferment. Here are these heirlooms that make amazing cider: I want to explore the tradition of what they might have been doing and that flavor profile."

Tasting Spiced and Botanical Ciders

TASTING SPICED AND BOTANICAL CIDERS

Apples and spice are natural complements; other foraged flavors add a unique twist to the traditional.

1. EMBER
Far from the Tree Cider, Salem, MA

What was once a winery dream became a cider reality when the couple behind this urban maker began fermenting apples instead. A fall release, the semisweet Ember is infused with toasted chai, smoked vanilla and burnt sugar to complement and add a spiced spark to the ripe apple base. 6.9% ABV

2. GOOD QUEEN BESS
Black Mountain Ciderworks and Meadery, Black Mountain, NC

Makers of Appalachian cider and mead, this husband-and-wife duo strives to make cider with natural components only. This sipper is a mix of cider and house-made lavender mead: a floral, rich, and off-dry cider fit for a queen. 6.5% ABV

3. FOREST GINGER
Finnriver Farm & Cidery, Chimacum, WA

Organic dessert and heirloom apples from Washington construct a sturdy, off-dry foundation for this cider that is fermented with fir tips harvested from Finnriver's property and with organic gingerroot. Simply put, it is sunshine and trees in a bottle. 6.5% ABV

4. FOR THE LOVE OF LILAC
WildCraft Cider Works, Eugene, OR

Nearly all production for this college-town cidery is done through wild fermentation with wild or abandoned fruit, flowers and herbs. Community-sourced lilac blossoms are fermented into a wine and blended into a dry cider base for a unique cider of maximum fragrance and flavor. 6.5% ABV

5. GIN BOTANICAL
Seattle Cider Company, Seattle, WA

Seattle's first cidery since Prohibition sources spent gin botanicals from a nearby distillery and blends them into a semi-dry, dessert fruit cider that allows the herbs to shine. Beaming with citrus, juniper, cardamom, and verbena, this aromatic cider asks for like-minded herbed roast chicken. 6.5% ABV

6. POME MEL
Colorado Cider Company, Denver, CO

Pome for the fruit and *mel* for the honey, this production team adds rosemary, lavender, and honey to its Colorado-grown Granny Smith apples for a cider that is forward in wildflowers, resin, honey, and pure apple flavors. 6.5% ABV

7. CUVÉE WINTERRUPTION

J.K.'s Farmhouse Ciders, Flushing, MI

An answer to the cider call of winter, this cider from the organic farmhouse cidery uses cinnamon, vanilla, and a skosh of maple syrup from trees on the farm. The result is unfiltered, roasty, toasty, and spiced, with plenty of spotlight to share without overwhelming the estate fruit. 6.9% ABV

8. ELDERFLOWER

Angry Orchard Hard Cider, Cincinnati, OH

From the seasonal cider lineup, this bevvie is available throughout spring and summer and bottles the aromas of the blossoming season. Elderflower and apple dominate the aromatics, while the ripe, bittersweet palate is toned with lychee and citrus. Considering the florals, this cider could take on vegetables or salads with ease. 5% ABV

9. GINGER

Eaglemount Wine and Cider, Port Townsend, WA

This peninsula cidery proudly lays claim to the first commercially produced ginger cider, using organic fresh gingerroot in its organic heirloom apple cider ferment. Semisweet, spicy, and refreshing, the cider can soothe and synchronize with flavors of Asian fare. 8% ABV

10. FRUIT OF FOREST

Nine Pin Ciderworks, Albany, NY

The first cidery in the state to receive a "farm cidery" license, Nine Pin is required and obligated to ferment only New York fruit. They do so proudly, even in their more avant-garde ciders like this sipper, an apple, sour cherry, and staghorn sumac cones coferment. The foraged cones bring a woody minerality to the powerful tang of the cherry and apple. 6.4% ABV

11. SUICIDER

Bishop Cider Co., Dallas, TX

This is a grown-up homage to childhood soda-fountain flavor mixtures, but with apples and baking spices like clove, allspice, and cinnamon. Intrinsically linking images of apple pie, the semisweet cider is crisp, flavorsome, and warming with its natural seasonings. 6.5% ABV

SPECIALTY CIDERS

Ask and you shall receive: cider drinkers across North America have demanded for what is next in the world of cider, with many drinkers coming from the beer world's focus on constantly refreshing novelty. Outside of spices, botanicals, and other fruit additions, some cidermakers set their sights on similar visions to their brewing comrades, tasting and testing from a wide variety of adjuncts they can add to their apple base — from coffee beans and habanero peppers to ciders fermented with wine grape pomace, Belgian ale yeast, and squash. Get ready, world, here they come: the catchall category of specialty ciders is big and gaining speed.

THE HISTORY

"Specialty cider" is a modern marvel; it's a comprehensive term that encompasses a variety of styles that change along with the enterprise, innovation, and extremity of the maker. The Beer Judge Certification Program (BJCP) categorizes ciders and defines specialty cider and perry as "beverages made with added flavorings (spices and/or other fruits), those made with substantial amounts of sugar-sources to increase starting gravities. . . ." The all-encompassing classification opens the door for many interpretations, but it does require the addition of something beyond apple (or pear) and yeast.

The heralded Great Lakes International Cider and Perry Competition (affectionately known as GLINTCAP) defines the category similarly. For this respected annual competition, a specialty cider or perry is an "open-ended" variation "with other ingredients such that it does not fit any of the [other] categories" like wood-aged, fruit, or even hopped/herbal cider.

Many of the makers dominating this category are located on the wild West Coast, where interpretations of cider run rampant, fermenting with ale and Belgian yeast strains and blending in everything from rhubarb, peppers, and carrot juice into ciders. Other makers around North America have caught on and started experimenting with flavors they've enjoyed elsewhere — in a beer or even from an Asian foods store. These ciders stand out on the shelf not because of a familiar name or meticulously designed label but because of creativity inside the bottle.

THE BREAKDOWN

In a perfect world, complex cider fruit or heirloom apples would be, well, growing on trees — everywhere. But the reality is that these specific varieties aren't widely available, and using the fruits of the behemoth culinary and dessert fruit market is the only option for a large number of cider producers. Taking this simple template and designing a cider with a more sophisticated flavor can require some inventiveness.

Brad Page of Colorado Cider Company says they knew they had to do something different from the beginning, considering the burgeoning market, the fruit crop of his home state, and his large source of apples from the Pacific Northwest.

"There are so many products on the market, shelf space is so important, and trying to get all of your products on it is daunting," Page says, who produces a lemongrass and hop–infused cider, along with other botanical and fruit-flavored ciders. "We approach specialty cider from a standpoint of not trying to make it like beer; we want it to have its own identity."

What makes the cider blow its
   cork with such a merry din?
What makes those little bubbles
   rise and dance like harlequin?
It is the fatal apple, boys, the
   fruit of human sin.

— Christopher Morley, "A Glee upon Cider"

Head Radical

Nat West, Reverend Nat's Hard Cider, Portland, Oregon

When it comes to the ensemble of personalities in the cider industry, the collection is a cast of characters. Opinions are unbridled, schools of thought vary not only by region and generation but by formality of education, sales success can trump quality, and the subjectivity of taste is often debated.

The gospel according to Nat West, better known as Reverend Nat West of his eponymous Portland, Oregon, cidery, is a viewpoint announced loud and clear among the rest.

"If you make something that is unique, if you make an edible consumable product that is strongly flavored and very innovative, you're bound to piss off some people, and others will love it."

Flagged by his strawberry blond, coarse goatee and often clad in his own logowear, West is a man respected by many for his innovation in flavors, applauded for his voice on modern cider definitions, and barbed by some for his avant-garde production tactics.

"As years go by, tastes change. If innovators had said, 'Let's make what everyone's making, let's make what everyone's drinking,' then we wouldn't have this development of taste, development of style in any of our food culture," West says. "It takes people who really love to drink and consume and eat certain flavors doing it again and again, and for other people to notice."

Reverend Nat's Hard Cider rotates through an ample selection of creative ciders, experimenting with flavors like pineapple, watermelon, ginger, lemongrass, hops, apricot, Mexican piloncillo negro sugar, sour cherry, fermented carrot . . . The list goes on, and West doesn't plan to stop.

"I could certainly just ferment apple juice and put it in a can with a pretty label on it and make more money, but that's not why I wake up in the morning," he says. "I'm not interested in making something that's somewhat milquetoast. I want to make something really unique and unusual."

West admits the first cider he ever tried was his own. After making cider in his basement from a friend's backyard tree for a handful of years, he took his personal supply commercial in 2011, selling in two months all that he had made. This validation encouraged him to expand from his home and into a 3,000-square-foot space the cidery now calls home, which has gradually expanded to two-and-a-half times its size.

With his devout beer-drinking palate, West has modeled his cider after the market he knows so well as a consumer, and he steers clear of attempting an orchard of his own.

"Beer guys don't sit around complaining about how the harvest was low this year or how the grain didn't get enough water or how the hops weren't south-facing or the terroir

> **"**
> As years go by, tastes change. If innovators had said, 'Let's make what everyone's making, let's make what everyone's drinking,' then we wouldn't have this development of taste, development of style in any of our food culture."

of the barley field," he says. "They all have access to the exact same ingredients, and yet if you give a pile of brewing ingredients to one brewer and the same pile to another, those beers are going to taste completely different depending on how they make the beer. That's how we approach our cidermaking. How I like to make ciders is more of 'a hand of the brewer' than 'hand of God.'"

Although his apples are 100 percent Northwest, the list of Reverend Nat's ingredient sources site on a map like an international tour. "We source from around the world for the best tasting of whatever the ingredient is that we're looking for," West says. "I'm going to wait for the best, not just any variety but the same variety I buy from the same producer every year."

He likens his ingredient selection to that of a chef — emphasizing that a cook wouldn't find himself in a field trying to grow each and every produce item he needed for his kitchen; he would find the top producer to get what he needed. This notion goes back to why West says he doesn't make traditional ciders, why he chose to experiment with flavors. Although he doesn't have an orchard, he finds that his ciders are an expression of the market rather than the terroir of the Pacific Northwest.

That market is calling, and Nat West is answering. As evidence, he reminiscences about the time international cider legend Tom Oliver, of Oliver's Cider and Perry in Britain's

Herefordshire cider region, introduced his first hopped cider at the annual Reverend Nat's Hopped Cider Fest. According to West, Oliver's go at the variation was dim, but the luminary's intent was brighter. "The guy just shows up to things and wins awards and he's embracing innovation," West says. "If that's not proof that cider is a malleable, adaptable thing that works with the ingredients at hand, then I don't know what is."

Apples are pressed for a second time through a horizontal press, affectionately known as a "squeeze-box."

Craft Spirit

Dave Takush, 2 Towns Ciderhouse, Corvallis, Oregon

Many American dreams have started in the garage. Google, Hewlett-Packard, Amazon, and Disney might ring a bell for success stories that were hurled into blockbuster orbit from the start-up headquarters of a carport. For childhood friends Dave Takush, Lee Larsen, and Aaron Sarnoff-Wood, their first batch of cider was made in an Oregon garage and tested at Larsen's wedding reception. The thrilled response from the guests was all the trio needed to take their plan to the next level of starting a cidery.

"We had a two-car garage, built the walk-in coolers together, dug a computer out of someone's trash can and repaired it, and our delivery vehicle was a 1996 Nissan Altima," Takush laughs as he details the team's humble beginnings in the 900-square-foot space. "We pulled our life savings together, which as three 20-somethings was basically nothing. It was a true bootstrap operation, and we thought if we can get this shoestring facility up and running and crank out cider at full capacity, within two years we'll consider it a success."

Within two months of their first commercial release, 2 Towns Ciderhouse had maxed out of the small garage's volume, with the two founding towns that inspired the cidery name, Corvallis and Eugene, Oregon, drinking more of their cider than they could produce. According to recent

data, 2 Towns Ciderhouse now is the fourth-largest independently owned producer in the United States.

"We are becoming a fairly big player. In terms of the cider industry and volume, we're growing a lot," Takush says. "Because of this, we are actively reaching out to local farmers, growers, and processors around the Northwest because we refuse to add fruit concentrates to our ciders. We really strongly believe that whole fruit is the way to go."

For 2 Towns' expansive cider line, the offerings vary from single-orchard select heirloom fruit bottlings to marionberry, rhubarb, cherry, hops, spices, ginger, and more infused into the apple blends. Knowing apple is the priority in their ciders, Takush says ensuring the use of these whole-food specialty items locks in more than just flavors of the adjuncts: the consumer gets the whole experience of the cider. "We're really stoked, because even though we're growing, we're looking into the future so we can grow in a sustainable way without cutting quality," Takush says.

Proponents of "what's next," some of 2 Towns' greatest hits have been their left-field ciders like those with rhubarb, hops, baking spices, bizarre and foreign fruits, or even the Ginja Ninja, a ginger-infused cider and ode to Takush's redheaded nature.

Much of the fruit for 2 Towns is coming from Washington State and Oregon's Hood River Valley, but in the first year of production, the team planted three acres of French and English cider varieties that Takush manages on his parents' farm. He recognizes that although the majority of their fruit comes from elsewhere and many of the ciders are highly experimental, orcharding has given him a lot more respect for the fruit.

"I'm very much primed in fermentation science," Takush says, who received his bachelor's degree in this field and his master's in food science from Oregon State University. "But the orchard has given our company balance, being connected with actually growing fruit. The same thing goes for when we source from local farms: it gives you a new perspective on this industry in general."

2 Towns Ciderhouse strives to straddle both worlds of the cidermaking mind-set. "With our ciders, we have a vast array from more traditional and eclectic, old-world ciders all the way up through some pretty wacky stuff that is really pushing the boundary of what cider is and can be, just like in craft beer," he says. "We like to say we have a winery license but a craft beer spirit."

> " We have a vast array, from more traditional and eclectic, old-world ciders all the way up through some pretty wacky stuff that is really pushing the boundary of what cider is and can be, just like in craft beer."

Tasting Specialty Ciders

TASTING SPECIALTY CIDERS

And then there are the ciders that nearly defy classification. Specialty ciders include a variety of flavors, from coffee beans to lemongrass.

1. TOTALLY ROASTED

Vander Mill, Spring Lake and Grand Rapids, MI

With two locations, this Michigan cidery ramps up their cider line with orchard-driven sippers alongside unique blends and infusions like this cider spiked with cinnamon, pecans, and vanilla. Spicy, nutty, and juicy with fruit, this cider could easily replace a morning coffee. 6.8% ABV

2. DELIVERANCE GINGER TONIC

Reverend Nat's Hard Cider, Portland, OR

This sipper compiles the spice from gallons of pure squeezed gingerroot juice with hand-cut lemongrass, hand-extracted quinine from Peruvian cinchona tree bark and fresh citrus juice to an off-dry dessert apple blend. The result has been a cider cult-classic, promoting the natural health benefits of all ingredients involved. 6.1% ABV

3. RHUBARBARIAN

2 Towns Ciderhouse, Corvallis, OR

Not quite a fruit and technically a vegetable, rhubarb stalks are pressed and gallons of juice go into this dry, English-style cider from the rapidly growing cidery. Far from the familiar pie, the aromas of this cider are tart, tangy, and somewhat feral and funky, giving new meaning to the ambiguous plant. 5% ABV

4. THREE PEPPER

Seattle Cider Company, Seattle, WA

Freshly sliced habanero, poblano, and jalapeño peppers are added to the fermenting juice and finish with a kick to the face. Must like spice, but the gracious dessert apple fruit soothes with an off-dry fruit-forward flavor that is secondary to the aromatic profile. Calls for guacamole. 6.9% ABV

5. ACE PUMPKIN

California Cider Company, Sebastopol, CA

Thanksgiving in a glass, this popular autumnal cider blends in pumpkin, allspice, cinnamon, and cloves after the juice ferments. Straightforward squash and baking spices sit semisweet until leisurely brisk acid sweeps up the sweetness. Proudly the West Coast's first pumpkin cider. 5% ABV

6. LEMONGRASS

Sietsema Orchards and Cider Mill, Ada, MI

Tenured apple orchard and U-pick farm in Michigan, Sietsema — and its cidery line — keeps to the basics but steps out every once in a while, like with this award-winning, lemongrass-infused cider. Zesty, citrusy, tart, and earthy, a semisweet cider base allows the zippy plant to shine. 6.9% ABV

7. HOT CHA CHA CHA
Woodchuck Hard Cider, Middlebury, VT
From the cidery's "out on a limb" series, this cider is fermented from bittersweet apples and is infused with Bird's Eye chili peppers. Heated, spiced, and ornery, this medium-bodied cider is not for the faint of heart but is for meals of burritos. 5.5% ABV

9. CIDER NOIR
Texas Keeper Cider, Austin, TX
Northern Spy apples are double-fermented with extra dark Belgian candi (a marriage of beet and date sugars), pecans, and orange peel, and then aged in oak barrels. These classic components of Belgian ales might be new to cider but are not quickly forgotten — the flavorful combination resonates and lingers for prolonged pondering. 10.1% ABV

9. FOUNDERS RESERVE ROSÉ
1911 Hard Ciders, Lafayette, NY
Taking the best of both agricultural worlds in upstate New York, this off-dry cider is blended with hybrid wine grapes, Noiret and Chancellor. Granting the glint of pink, the grapes also offer the juicy pomegranate and berry flavors, while the fermented apple stabilizes with acid and yeast components. 6.9% ABV

10. HIBISCUS SAISON
Common Cider Company, Drytown, CA
Tropical, trumpet-shaped, and purportedly containing healing qualities, flavor-enhancing hibiscus is blended into this semisweet cider blend. Fermented with saison yeasts, wheat beer flavors and textures add creaminess to this floral, food-friendly cider. 6.5% ABV

PERRY

Hendre Huffcap, Bloody Bastard, Stinking Bishop, Mumblehead — perry pears do not have the most alluring monikers. Despite this, makers of perry are trying hard to turn it into a drink that is as popular as cider. Because of a general lack of awareness, perry — the fermented juice of pears — is often confused with *pear cider*, which is cider with pear juice in it. According to many who produce from the fickle fruit, there is a lack of labeling and packaging distinction between the two, which results in an even more lost consumer. Perry, the delicate, earthy, and Champagne-esque tipple beloved by Europeans across time, is just starting to flourish in the cider market, and for good reason.

THE HISTORY

It might have been fourth-century Gallo-Roman theologian Palladius who said that Romans preferred "wine made of pears" to that from apples (and provided details on how to ferment the juice), but it was Frankish emperor Charlemagne who was the first on record to label fermented pear juice as *pyratium*, Latin for perry, in the ninth century. Charlemagne also takes the credit for giving cider (*pomatium*) a name, and his power was felt on many terrains throughout Spain, France, and Germany, with cider and perry moving with the merging cultures. The eleventh-century Norman Conquest of England brought pears and perry to where it is most popular today,

particularly in the West Country's oceanic climate and mineral-rich, fertile soils, which includes cider homes Somerset, Dorset, and Herefordshire.

Perry pear trees age slowly and have the ability to bear fruit for centuries. They also grow to towering heights; because of this, the trees were historically used as shelter for apple orchards against the harsh winds of the English seaboard.

Traditional perry production, similar to that of cider, requests perry pears but does not require it. Perry pears are thought to be hybrids of wild pears and the cultivated pear (*Pyrus communis*). Like bittersharp cider apples, perry pears are high in tannin and acid, inedible to a more discerning palate but ideal for fermentation.

THE BREAKDOWN

Growing enough perry pears might be the biggest hurdle to producing perry in North America. There aren't many of these trees on the continent, and importing rootstock from abroad is difficult. On top of that, it can take anywhere from five to ten years for the trees to produce fruit. Even then, the success of growing perry pear trees and its fermentable fruit is inconsistent, not to mention harvesting and juicing the fruit itself is extremely labor intensive.

Pears also require special handling. Compared to apples, pears (perry and culinary) have significantly divergent chemical and organic compositions. For starters, pears are higher in citric acid, whereas apples are made up of mostly malic acid; the former can quickly convert to acetic acid (like in vinegar) if not handled carefully. Second, pears contain a considerable quantity of sorbitol (unfermentable sugar) that provides body to the drink as well as residual sweetness postfermentation. Too much body, or texture, can lead to a syrupy consistency; thus many makers choose to add carbonation to avoid any offense.

As growers of cider apple trees know, terroir makes a difference in cider flavor. When fermented, apples grown in North America taste quite different from the same variety grown in its ancestral locations of England and France. The aroma, flavor, and textural profiles can differ from what the maker or grower might read in a textbook. The same is true for pears.

Although you might think all this would be a deterrent to perry makers, these challenges have actually inspired producers to experiment, including reaching beyond traditional perry pears and using accessible domestic fruit like Bartletts, Bosc, Seckel, and Comice. The search for the intricate, soft, and wispy aromas and flavors of perry can come through regardless of the pear; much of it depends on the treatment of the fruit and the gentle hand of the maker.

Professor Perry

Charles McGonegal, ÆppelTreow Winery & Distillery, Burlington, Wisconsin

When Charles McGonegal gets going, his cadence slightly mirrors that of actor Paul Giamatti from 2004's "enocentric" flick *Sideways*. Although his tone avoids the irate intensity often seen in Giamatti's characters, he skips along with a similar syncopation to his discourse about fermented fruit beverages.

"I'm an orchard-driven cidermaker for 90 percent of what I do," McGonegal says when talking about the current cider and perry scene. "I respond to my locals somewhat, but I will continue to show what the orchard and land does, because that to me is being true to the apple. That is our name, it's written in Anglo-Saxon. Nobody can pronounce it, but the philosophy is that we're true to the apple."

The name, ÆppelTreow (pronounced "apple true") Winery & Distillery, is something McGonegal now writes off as a marketing fail that stuck. But he isn't a branding mogul, he is a man of science. Educated as a biochemist who spent his career in petrochemical research, McGonegal says there are a few who call him "the Professor" in the cider and perry industry — because he approaches cidermaking from a scientific perspective — and it's a title he holds with pride.

He began making cider under his dorm room desk while at Michigan Technological University and caught the fever for fermenting, a desire that sent him on a mission for specific varieties that would excel during the science project. While on his hunt, he found Brightonwoods Orchard in Burlington, Wisconsin, home to 120 heirloom cultivars, in 1997. He began to trade homemade cider for the heritage orchard fruit and, four years later, McGonegal and his wife, Milissa, moved their production into a 1930s dairy barn on the orchard's property. There, they were planting test blocks of fruit to trial their ciders with apples specifically for cider production, long before there was a demand for that type of fruit.

"I had already partnered with an orchard that had been there for 60 years. I was already that much ahead of the game with all these heirloom Russets and whatnot," McGonegal says, grateful for his fortunate situation. "They don't have the bitters, but they have 120 varieties to grow blends out of, all the American classics to learn from and to use."

For perry pears, his pick of the crop did not come as easily. After meeting award-winning maker Tom Oliver of Oliver's Cider and Perry in England's Herefordshire, McGonegal was introduced to perry pears, a fruit he describes as "mysterious, complex, and nasty" creatures that he had to try on his own. With the limited option of English

Co-owner Milissa McGonegal designs the artwork for the cidery's labels.

and Austrian perry pears he was able to obtain domestically, McGonegal accepted the challenge of acquiring French varieties and began the substantial process of importing samples, which required federal permits, state permits, a quarantine procedure that included genetic screening for known viruses, propagating and planting at a farm-laboratory until the tree fruits on-site, and more. Finally, the handful of French perry pear varieties he selected were released for provisional growing, and McGonegal is paying it forward.

"That's been a long adventure, and I still have things trickling through," McGonegal says, noting he made his findings public to the National Clonal Germplasm Repository, an agricultural research service agency. "I'm not trying to trademark them, I want people to have access, too. I think heirloom fruits are essentially open source.

Just because these are new [to the country], they are still heirloom."

McGonegal's battle with perry pears hasn't been easy — he once got in an argument over usage with the Chicago Botanic Garden regarding four neglected perry pear trees in their fruit collection — but it has put him on the map as an orchardist and maker of the drink.

"The Professor" says he likes to bring the science to the drawing board and connect the dots for others. True to his apples and pears, McGonegal says he is making a product that he is happy with and wants to educate the consumer with, as opposed to tracking the course of the market.

"I'm saying, 'Here's what I think is fine-flavored, orchard-driven cider; let's educate you and see if you agree with me,'" he explains. "I don't want to follow the market, I want to lead the market."

" I make an OK cider; I rarely claim to be the best at anything. But I make a good perry."

Pear Proprietor

Jim Gerlach, Nashi Orchards, Vashon, Washington

Nearly 6,000 miles stretch between the origin of *Pyrus pyrifolia*, the Asian pear, and its Pacific Northwest home on Vashon Island, a 37-square-mile islet just 30 minutes from downtown Seattle. This fruit flourishes in roughly 200 trees on 27 acres of Vashon's Nashi Orchards.

Nashi, which is Japanese for "pear," was an accidental endeavor for husband-and-wife duo Jim Gerlach and Cheryl Lubbert. "When we bought the property in 2005, we didn't have any aspirations of making perry. We didn't even know much about it at that time," Gerlach says of his agricultural mishap.

After renovating the rundown orchard on the property — which also held an Asian-inspired home with immaculate architecture — the couple realized they had way more fruit than they could give away. For Lubbert, an avid gardener with a microbiology degree, and Gerlach, a home brewer, gourmand, and landscaper, making lemonade from their lemons seemed the obvious solution.

"We needed to find a way to turn this liability into an asset," says Gerlach of the orchard that now sports Asian pear hybrids unique to the island. "We think these pears are fairly unique to us. They have characteristics that work really well for beverage making."

Contrasting traditional perry produced from the fair-weather European perry pear varieties (*Pyrus domestica*), Gerlach says Asian pears offer good acidity and earlier ripening times. In perry pears, of which he grows four different varieties, he says the pH can be too high and lead to spoilage issues, where the fruit ripens from the inside out, making the process even more arduous.

"Especially with pears, it takes quite a bit more finesse to make perry than cider," Gerlach says. "Perry can get pretty squirrely pretty easily, and you have to be fairly meticulous about making it. Asian pears are nice compared to European pears in that we can sit on them. We can let them mature fully and preripen as well, whereas European pears you pick them green, put them in a cold storage to ripen that way."

Perry has been Nashi's sole offering — captured in adorable 187-milliliter bottles donning the namesake fruit on the label — until a recent release of seasonal, small-batch ciders made from estate and island fruit.

Perry still takes priority in the orchard and in the cidery, a converted barn that houses both production and a modest tasting room open on weekends. With only 10,000 residents, Vashon Island opens itself to high tourism traffic, and Nashi's tasting room reaps the benefits of the funky farm life that draws in visitors. The couple introduces single-varietal perrys to cider lovers amid the orchard and scrupulously manicured gardens and between belly rubs with the three Bouvier des Flandres fluff-ball pups that run on the property.

"Perry fits really well into the idea of keeping everything on island," Gerlach says of his home. "We wanted it to be about this place."

Although Gerlach believes the Vashon cidery is in a "sweet spot" of both island consumers and the Seattle market, there are still plenty of obstacles perry and its makers have to jump over. "Even among perry makers, some of them call this 'pear cider,' and I say, 'You've got to be kidding me, is it made of all pears? Then call it perry,'" he laments, observing that other perry producers find the title a difficult sell over calling it "pear cider."

"It's your responsibility to educate the consumer," he tells his fellow makers. "That's one of those things we've found is that people are so interested and people are really excited about learning about things and finding out about how that's the difference between perry and pear cider. . . . It's not just about making a commodity product, it's about telling a story and giving people an experience they can really enjoy.

Gerlach says Asian pears offer good acidity and earlier ripening times than traditional perry pears.

Modern Orchard, Traditional Product

Marcus Robert, Tieton Cider Works, Yakima, Washington

Harmony Orchards has one of the best seats in the house. Perched at 2,000 feet above the convergence of the Tieton and Naches Rivers in Washington's Yakima Valley, a heralded fruit-growing region of the West, Harmony's fertile soils and resulting fruit enjoy long days and cool nights, a diurnal shift that makes the apples, pears, cherries, and apricots off the orchard unique to the region. Studied horticulturist and third-generation farmer Craig Campbell and his wife, Sharon, launched Tieton Cider Works off the organically grown, beautifully situated fruit of their Harmony Orchards in 2008, planting cider apple varieties that same year under the calculated guidance of business partner and cidermaker Marcus Robert.

Nearly a decade later, Tieton has a second brand, Rambling Route, a new and trending cider bar in downtown Yakima, Washington, and the largest acreage of cider apples and perry pears on the West Coast.

With a few hundred cases of production for the first harvest in 2008, Tieton produces nearly 100,000 cases today from more than 400 acres of high-density orchard structuring. Roughly 2,500 trees are planted per acre, in comparison to conventional orchard plantings that grow about 700 trees per acre. The upshot is the model lends itself to sooner returns on crop — fruit is growing faster on trees that typically take years upon years to fruit.

"All of the testing we've done in the past on apples has been to grow them in our climate," Robert says, a fourth-generation orchardist himself. "The research lent itself to a modern orchard like this. Really nobody else was doing the work on all those given critical points to produce cider apples in our climate."

Nobody else was or is growing what Tieton is growing, particularly with the intent of cider on this scale. This commitment to the land, to the fruit that comes from it, is something that is of the upmost importance to Robert's production.

"I think it's very critical that people understand that even if you're not an orchardist, you understand the orchard operation and the background there," he says. "We find that people come to us and maybe want apples, want different things done, but the seasonality, the propagation, and orchard management aren't well understood by a lot of people that come into the industry."

Although the mind-set might be modern, Robert's cider and perry production is influenced greatly by European makers. "We planted perry pears because we enjoyed the beverages that were being imported here like Oliver's, Bordelet, DuPont," Robert says. "We rounded up some perry wood, planted a test block, and now we have six acres of perry pears which is probably the largest block in the United States."

Using 100 percent perry pears — with fantastically hideous names like Butt and Gelbmostler — Tieton's perry is made from fruit that is high in citric acid and tannins and are small, not lending themselves to much juice nor an easy process. Using the traditional French cider method of keeving, a process in which natural sweetness and carbonation occurs from stopping the fermentation before dryness, Robert's chosen process allows him to bottle an elevated product unlike many others made on the continent.

"We are really taking a traditional product and we're taking these pears that are used to produce one product — perry," he explains. "We're looking at the orchard, bringing it to the table, talking about heritage. The perry pears we are using have hundreds of years, if not thousands of years, of tradition and skill behind making it."

Using 100 percent perry pears – with fantastically hideous names like Butt and Gelbmostler – Tieton's perry is made from fruit that is high in citric acid and tannins and are small, not lending themselves to much juice nor an easy process.

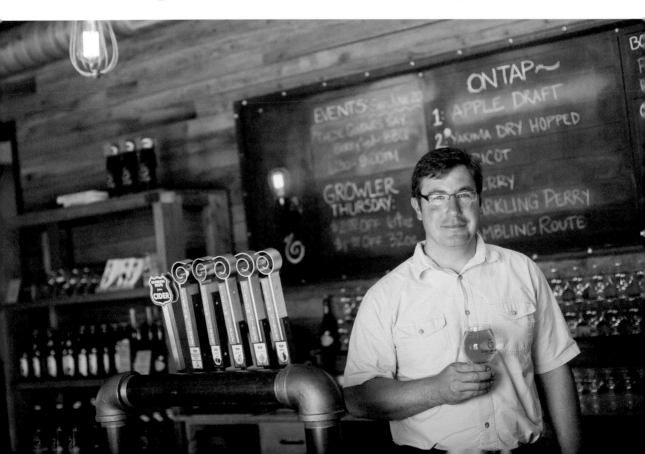

Tasting Perry

TASTING PERRY

Whether it's labeled as pear cider or perry, these fermented pear selections are simply bursting with delicate floral pear flavors.

1. PERRY

ÆppelTreow Winery & Distillery, Burlington, WI

This maker has gone through the trials and tribulations of importing French perry pears, and it's proven supremely worthwhile, as this applauded perry sets the bar for traditionally produced perrys in North America. Delightful and dainty, the semisweet drink is its best self when next to Brie cheese. 8% ABV

2. CHOJURO BLEND ASIAN PEAR PERRY

Nashi Orchards, Vashon, WA

Island-grown Asian pears are fermented in a miniature cider house in even smaller equipment for a perry of big flavor and texture. Maintaining delicacy, this perry speaks to its juicy base fruit, showing layers of blossoms, soft stone fruit, and accessible acid. A nearly unbeatable pairing with scallops. 6.7% ABV

3. SPARKLING PERRY

Tieton Cider Works, Yakima, WA

Arguably the largest perry pear (and European cider fruit) grower on the West Coast, Tieton bottles traditional perry made from those European perry pears. These varieties fuel this Champagne-esque dry perry, a dazzling, mineral-rich, and pleasantly lower-alcohol corked offering. 5.5% ABV

4. PERRY

Pommies Cider Co., Caledon, ON

All pear, all the time — Ontario-grown Bosc and Barlett, to be specific. Aromatics of white flowers, spring pear, and sweet pea fill the glass, while sprightly and easy bubbles lift up the soft and sweet pear fruit flavors. Easy drinking and easy pairing, try it with ceviche. 5% ABV

5. PERRY

Sea Cider Farm and Ciderhouse, Saanichton, BC

When a secret garden of perry pears were revealed to cidermaker Kristen Needham, she jumped at the chance to make traditional perry. Years later, she still has access to this fruit, producing a perry that is elegant, fragrant, and lustrous in island-generated acidity. 6.5% ABV

6. PEAR CIDER

Island Orchard Cider, Ellison Bay, WI

Francophiles by way of Wisconsin, this orchard-based cider house channels Normandy when making its ciders and perry. The latter balances Bosc and Barlett pears in a blend that is lush in pear fruit, mineral, and spice yet light-footed and bouncy. 6.8% ABV

7. POIRE
E.Z. Orchards, Salem, OR
Traditional French methods are used for this finely textured and sharply tuned perry from Oregon wine country. Estate-grown Forelle, Comice, and Bosc winter pears make up the composition and spread wide with a creamy palate that finishes crisp and clean. Lighter alcohol allows for multiple servings. 4.8% ABV

8. PERRY
Rootstock Ciderworks, Williamson, NY
Multigenerational DeFisher Fruit Farms first launched a distillery focused on apple spirits before starting this cidery, which now bottles half a dozen regular bevvies. The perry is 100 percent Bartlett pears from the farm and shows that with slight funk touched with flowers, acidity, and true pear fruit. 6.7% ABV

9. PEAR ESSENTIALS
Neigel Vintners Cider, East Wenatchee, WA
Hopelessly devoted to pears, the brothers behind this central Washington orchard came into fermenting the fruit when they came into ownership of their grandfather's pear estate. This perry is the top-of-the-line offering from Neigel, juicy, rich, and bold with pear fruit and tangy acid. 5.2% ABV

10. PIONEER PERRY
WildCraft Cider Works, Eugene, OR
The signature (and vintage) perry from this wild-ferment and foraged cidery uses American pear varietals that rooted in Oregon before Prohibition. Complex and compelling, this feral, unpasteurized, and bottle-conditioned sipper is delicious with breakfast. 6.9% ABV

SHAKEN, STIRRED,

AND

SERVED UP

CIDER COCKTAIL RECIPES

No longer is cider exclusively associated with fall or even autumnal imbibing. Mixologists, bartenders, and cocktail designers across North America are adding and emphasizing cider in recipes, season to season, flavor by flavor. In this compilation of cider's greatest cocktail hits, bartenders share their secrets for using the fermented fruit in heady concoctions, ripe for at-home mixing. Step outside of your booze box: some recipes require a trip to the liquor store, while others provide directions on house-made syrups and tinctures to enrich your home bar. Specific brands noted are the preferred choices of the bartenders who created the recipes; feel free to substitute with similar spirits and ciders that are local to your region.

CIDER IN THE CLEAR

GIN, VODKA, AND AQUAVIT

Far from a blank slate, clear spirits like gin and vodka provide complementary flavors, textures, and complexities to a cocktail, with the added bonus of fruit, acid, and tannin that cider can offer. Toss in the wild card of aquavit, the Scandinavian spirit flavored with herbs and spices, and you've got yourself a full deck of cider cocktails ready to be dealt.

Eldermint Lavender Bloom

Recipe by Kate Bigham, Capitol Cider, Seattle, WA

In Seattle, the shoulder season between the holidays and spring can be a slow, dreary drag. As a solution, Kate Bigham serves up this "transitional cocktail" to guests at the dimly lit and rustic cider outpost she manages in the food and drink–concentrated Capitol Hill neighborhood. Her combination of the bright red raspberries and floral elderflower liqueur with the savory, wintery mint and off-dry bitterness of the Wandering Aengus Ciderworks Bloom cider helps make that seasonal transition easier. Spiked with gin distilled from Washington State apples and local botanicals, the aromatics alone carry imbibers into the spring season.

DIRECTIONS

Makes 1 cocktail

Combine the gin, elderflower liqueur, raspberries, and cider in a shaker and fill with ice. Shake vigorously; do not strain. Pour into a collins glass. Top with the bitters and garnish with the mint.

INGREDIENTS

- 1 ounce BelleWood Distilling gin
- ¾ ounce St-Germain elderflower liqueur
- 3 raspberries
- 1 ounce Wandering Aengus Ciderworks Bloom cider
- 1 strong dash of lavender bitters
- 2 mint sprigs, for garnish

THE CIDER

BLOOM

from Wandering Aengus Ciderworks, Salem, OR
A blend of heirloom apples found in both Salem and Hood River, Oregon, Bloom is an off-dry sipper that falls in line with the spring-blossom aromatics of the elderflower liqueur and lends an earthiness that strides alongside the mint and tart berry.

Poinciana

Recipe by Tim Prendergast, ANXO, Washington, DC

As of spring 2016, America's capitol city has its own cider bar and cidery in one. Boasting Basque-inspired small plates, or *pintxos* (pronounced "peen-chos"), the bar promotes one of the country's largest cider menus, soliciting not only international gems but their own house-made ciders as well. According to ANXO beverage director and certified cicerone Tim Prendergast, the Poinciana is his take on a classic French 75 Champagne cocktail comprising gin, sugar, lemon, and bubbles, subbing Millstone Cellars' traditional farmhouse cider for the sparkling wine and adding two of his favorite ingredients: Cynar (bittersweet artichoke liqueur) and Angostura bitters. "It's a drink that checks many boxes," Prendergast says. "It's strong, it's earthy, and it's complex, while still retaining a thirst-quenching lightness."

DIRECTIONS

Makes 1 cocktail

Combine the gin, Cynar, lemon juice, syrup, and bitters in a shaker. Top with ice, shake, and strain twice (to remove all ice chips) into a large, chilled rocks glass. Top with the cider and garnish with the lemon peel and cherry.

INGREDIENTS

- 1½ ounces gin
- ¼ ounce Cynar
- ¾ ounce lemon juice
- ¾ ounce simple syrup
- 1 dash of Angostura bitters
- 3 ounces Millstone Cellars Farmgate cider Lemon peel and cherry, for garnish

THE CIDER

FARMGATE

from Millstone Cellars, Monkton, MD

Tucked into a rural Baltimore suburb, Millstone Cellars is based in a restored historic mill and thrives in its rustic surroundings to produce New England farmhouse ciders. The Farmgate is a "mainline" product for Millstone, a blend of three heirloom New England apples that brings both earth and a little funk to this cocktail.

Shacksbury Sling

Recipe by Sas Stewart, Stonecutter Spirits, Middlebury, VT

Stonecutter Spirits is all about barrel aging on another level — as in levels of barrels are stacked in their distillery and tasting room, with more than 300 casks and counting. The house-distilled gin is even tailored to fit the flavors of oak, and its botanicals also match the herbal Bénédictine and elderflower-flavored liqueurs of this drink. Drawing inspiration from the gin-based Singapore Sling, Stonecutter co-founder Sas Stewart updates this classic recipe to highlight both cider and gin. "The expressive fruit nature of the Shacksbury Farmhouse and the botanical and barrel notes from the aged gin balance the sweetness from the Bénédictine and elderflower liqueur, while the lime juice brightens the whole drink."

DIRECTIONS

Makes 1 cocktail

Combine the cider, gin, Bénédictine, elderflower liqueur, and lime juice in a mixing tin with ice and stir for 10 seconds. Strain into a tall collins glass filled with fresh ice. Clap the mint between your hands and garnish the glass with the bruised mint and a straw.

INGREDIENTS

- 2 ounces Shacksbury Cider Farmhouse cider
- 1 ounce Stonecutter Spirits Single Barrel gin
- ½ ounce Bénédictine
- ½ ounce elderflower liqueur
- ½ ounce lime juice
- Mint sprig, for garnish

THE CIDER

FARMHOUSE

from Shacksbury Cider, Vergennes, VT

This cider spans two continents: the Shacksbury team sources Jonagold, McIntosh, and Empire apples from a single orchard in Vermont, but also trips over to Herefordshire, England, to pluck English bittersweets from an orchard across the pond. A year-round staple for the innovative cidery, the Farmhouse sports a little funk with a little fresh via a little barrel aging, mirroring flavors of the aged gin and liqueurs it so cohesively mixes with.

King Crimson

Recipe by Jade Sotack, Wassail, New York, NY

New York's first full-fledged cider bar was opened in 2015 by well-curated and cultured cider enthusiasts and bar experts Jennifer Lim, Ben Sandler, and Sabine Hrechdakian, who also is a producer of Cider Week NYC. Natural light breaks through the windows of this cozy bar, with dark hardwoods and the scent of cider throughout that create an atmosphere reminiscent of an orchard. With a staff as culled as the cider list (an extensive and distinct offering of fermented fruit), head bartender Jade Sotack builds this cocktail to cap between the seasons, comprising bitter, sweet, and savory flavors.

DIRECTIONS

Makes 1 cocktail

Muddle the grapefruit in a shaker, then add the Cynar, gin, lemon juice, and grenadine. Shake and strain into a collins glass. Add ice and stir to combine. Top with the cider and garnish with the cranberries.

THE CIDER

OLD TIME

from South Hill Cider, Ithaca, NY

Upstate in Ithaca, New York, cidermaker, orchardist, and apple forager Steve Selin hunts for the ugly, the battered, and the bruised apples, sourcing much of his fruit from feral trees in abandoned orchards in the Finger Lakes. His Old Time cider is the culmination of wild heirloom apples and a handful of cultivated bittersweets. The off-dry cider reveals a glimpse of sweetness in tropical fruit and matches with the vibrant citrus and earthy Cynar in the King Crimson.

INGREDIENTS

- 2 grapefruit wedges
- ¾ ounce Cynar
- 1 ounce Peak Spirits Farm Distillery CapRock gin
- ¾ ounce lemon juice
- ½ ounce grenadine
- 3 ounces South Hill Cider Old Time cider
- 2–3 cranberries, for garnish

Raspberry Cider Collins

Recipe by Kathy Casey, Kathy Casey Liquid Kitchen, Seattle, WA

Internationally lauded mixologist and chef Kathy Casey prides herself on her knowledge of beverage, from scotch to shrubs and cider. As the cocktail columnist for our own CIDERCRAFT magazine, she also touts the title as the "original bar-chef," a handle she has worked hard to forge, melding her culinary background with her savvy behind the bar. For her Raspberry Cider Collins, Casey subs out soda water for raspberry cider that screams for a sunny day.

 DIRECTIONS

Makes 1 cocktail

Measure the vodka, lemon juice, syrup, mint, and thyme into a shaker. Fill the shaker with ice, cap, and shake vigorously. Strain into a tall glass and top with the cider. Garnish with the raspberry and lemon twist.

INGREDIENTS

- 1½ ounces vodka
- ¾ ounce lemon juice
- ½ ounce simple syrup
- 3–4 mint leaves
- 2 sprigs thyme
- 3 ounces Red Branch Cider Hard Raspberry cider, chilled
  Fresh raspberry and lemon peel twist, for garnish

THE CIDER

HARD RASPBERRY
from Red Branch Cider Company, Sunnyvale, CA
Inspired by traditional cyser — the ancient fermented blend of apple and honey — Red Branch Cider brings fragrant honey from the Pacific Northwest into the mix of its cider production. Fresh California raspberry juice is blended into the Hard Raspberry, an aromatic and pink-hued sipper that serves as sparkling water in this classic cocktail, enhanced with herbs and spiked with lemon.

Okay, Let's Talk about Magic

Recipe by David Greteman, Taste Bar, St. Louis, MO

Barman David Greteman falls for the fusion of pear and gin in this zesty cocktail. "I wanted a refreshing, all-weather sipper, and the dense profile of the gin seemed a great counter to this dry, crisp perry," Greteman says. The aforementioned perry, from Argus Cidery, is pure pear with a spike of gingerroot added and, according to Greteman, brings an assertive dryness to the sweeter ingredients in the drink.

DIRECTIONS

Makes 1 cocktail

Combine the gin, peach liqueur, lime juice, and syrup in a shaker. Shake well and strain into a highball glass with ice. Top with the perry.

Cinnamon Syrup

Makes 1 cup

- 1 cup water
- 1 cup sugar
- 4 cinnamon sticks, broken into small pieces

Bring the water, sugar, and cinnamon sticks to a boil in a saucepan, stirring to dissolve the sugar. Reduce the heat and simmer until the mixture is fully dissolved and slightly thickened, 8 to 10 minutes. Strain the syrup from sticks and store refrigerated until ready for use.

INGREDIENTS

- 2 ounces North Shore Distillery Distiller's Gin No. 11
- ½ ounce peach liqueur
- ¾ ounce lime juice
- ¼ ounce Cinnamon Syrup
- 2 ounces Argus Fermentables Ginger Perry

THE CIDER

FERMENTABLES GINGER PERRY

from Argus Cidery, Austin, TX

Argus Cidery scours the south for apples ideal for their fermentation, sourcing many from Texas and the maker's home state of Arkansas. The perry is a second addition to the cidery's "Fermentables" line of lower-alcohol drinks, also including a ciderkin, a historic beverage made by fermenting reconstituted apple pomace. Flavors of ginger, pear, citrus, stone fruit, and pear rise to the occasion for this cocktail.

Queen City Cobbler

Recipe by Stuart Jensen, Mercantile, Denver, CO

Mercantile is an amalgamation of farm and fine dining — a purveyor of fresh market goods from its private farm, a popular farm-to-table eatery, and an artistic libations bar. Bartender Stuart Jensen fuses the finesse of traditional cocktails with local ingredients in an old-school sherry cocktail with local cider from Denver's Stem Cider. "I wanted to make a cocktail that would highlight and accentuate the berry and tart apple notes in the cider without overpowering them," Jensen explains. "A slightly lower ABV cocktail seemed a good fit."

DIRECTIONS

Makes 1 cocktail

Muddle the cucumber, raspberries, and mint leaves in a shaker. Add the sloe gin, brandy, sherry, lemon juice, and syrup. Fill with ice and shake until well chilled. Strain into a collins glass with fresh ice and top with the cider. Garnish with the cucumber wheel, fresh raspberries, and mint sprig.

THE CIDER

BRANCH AND BRAMBLE

from Stem Ciders, Denver, CO

Enlightened about apples from their Michigan upbringings, Eric Foster and Phil Kao both came to Denver for reasons outside of cider but each was drawn to the fermented fruit and launched Stem Ciders. Its Branch and Bramble is a fully dry cider, merged with tart raspberry fruit, which Jensen says is a stellar stand-in for Champagne. "It has some tannin that lends balance to the cocktail, the carbonation makes the cocktail frighteningly drinkable, and the dry acidity makes all the flavors pop."

INGREDIENTS

3 cucumber slices
2 raspberries
6 mint leaves
¾ ounce Sipsmith Independent Distillers sloe gin
¾ ounce Campo de Encanto Grand & Noble pisco brandy
¾ ounce Lustau amontillado sherry
¾ ounce lemon juice
½ ounce simple syrup
2 ounces Stem Cider Branch and Bramble cider
Cucumber wheel, raspberries, and sprig mint, for garnish

Penicillin #3

Recipe by Nightbell, Asheville, NC

Under dim chandelier lighting and among art deco–inspired furnishings, the charming Nightbell brings contemporary cuisine and inventive cocktail creations to the people of picturesque Asheville. The bartenders focus on foraging local ingredients and robust spirits to splash into glasses with hand-carved ice and, in this case, spiked with cider moonshine. Nightbell provides the recipe for homemade apple pie moonshine in this cocktail to help you truly feel like a bootlegger in the South.

Makes 1 cocktail

Combine the moonshine, syrup, and lemon juice in a shaker. Add ice and strain into a rocks glass. Add one large ice cube and garnish with candied ginger, if desired.

Apple Pie Moonshine
Makes about 3 cups

15 ounces Noble Cider Standard Bearer cider
15 ounces apple juice
⅓ cup sugar
1 cinnamon stick
¾ cup and 2 tablespoons Everclear grain alcohol

In a saucepan, bring the cider, apple juice, sugar, and cinnamon to a boil. Let cool, then stir in the Everclear. Refrigerate until ready to use.

Honey-Ginger Syrup
Makes about ¾ cup

10 ounces peeled fresh ginger, cut into thumb-size pieces
1 cup boiling water
¾ cup honey

Place the ginger and boiling water in a blender and combine thoroughly. Strain the mixture through a coffee filter into a bowl. Add the honey while the mixture is still hot, and stir until the honey dissolves. Let cool and refrigerate until ready to use.

2 ounces Apple Pie Moonshine
¾ ounce Honey-Ginger Syrup
¾ ounce lemon juice
Candied ginger, for garnish (optional)

STANDARD BEARER
from Noble Cider, Asheville, NC
Noble Cider makes proud Appalachian ciders with fruit from the western side of the state, poured like water from its taproom in the booming beer city. The Standard Bearer is the cidery's flagship product, a clean, dry "everyday, everybody" cider for the masses. Boiled down to its most simple characteristics, the true apple flavor shines in the homemade hooch alongside the intensity of the Honey-Ginger Syrup.

Ragnar's Ruin

Recipe by Mattias Hägglund, Heritage, Richmond, VA

Cofounder and mastermind behind the bar, Heritage's Mattias Hägglund embraces his home state cider scene by supporting local ciders on draft and shaken up in crafty cocktails. "Approach just about any bartender in Richmond and ask what their desert-island cider would be, and most will tell you Potter's Hop Cider," Hägglund says. "In this cocktail, the Citra and Amarillo hops provide a great backbone to the tart, dry cider. It balances the creamy minerality of the oyster shell–infused aquavit. And with the addition of floral absinthe, a touch of juniper, and a little lime and sugar, this becomes a really refreshing sipper."

DIRECTIONS

Makes 1 cocktail

In a shaker, combine the aquavit, lime juice, syrup, absinthe, and tincture. Add a small amount of ice and briefly shake. Double-strain into a collins glass, fill with fresh ice, and top with the cider. Garnish with the dried apple chip and mint sprig.

Juniper Tincture
Makes about 1 cup

- 4 ounces freshly crushed juniper berries
- 8 ounces 100-proof vodka

Combine the berries and vodka in a mason jar. Attach the lid and shake to mix. Store in a cool, dark cabinet for a week, then strain through a fine-mesh strainer into a bottle.

INGREDIENTS

- 1½ ounces James River Distillery Øster Vit aquavit
- ¾ ounce lime juice
- ¾ ounce simple syrup
- 2 dashes absinthe
- 12–15 drops Juniper Tincture
- 3 ounces Potter's Craft Cider Hop cider
  Dried apple chip and sprig mint, for garnish

THE CIDER

HOP CIDER
from Potter's Craft Cider, Free Union, VA

This small-town cidery uses vintage Virginia fruit — like GoldRush and Albemarle (Newtown) Pippin — as the base of its Hop Cider, the first in its series of hopped ciders. Bitter and citrusy as it should be with the use of Citra and Amarillo hops, this cider matches the botanical flavors of the juniper tincture while melding with the fennel of the absinthe and the spices of the aquavit.

CIDER SOUTH OF THE BORDER

TEQUILA, MEZCAL, AND RUM

Leave the lime and salt at home: cider brings a new flavor to the table with these cocktails that accentuate the sugar-based spirits of Latin America.

Innovative and creative bartenders fuse the top ciders of their region with the piquancy of tequila, mezcal, and rum for the ultimate in cross-country imbibing.

El Negro Amargo

Recipe by 101 Cider House, Westlake Village, CA

The first cidery to open in the City of Angels, 101 Cider House brings local to Los Angeles, courtesy of two Angelenos. All of the ingredients used in the ciders are sourced from orchardists, botanists, and farmers along U.S. Route 101, the highway that runs from Los Angeles to Olympia, Washington. True to its Southern California roots, this house-made cocktail pulls in tequila from the south with the Black Dog cider and laces it with homemade lavender syrup and a spritz of citrus to create the tart and refreshing drink.

DIRECTIONS

Makes 1 cocktail

1. Rub the rim of a rocks glass with one of the lemon slices. Place the sugar on a small plate and place the rim into the sugar. The sugar will adhere to the moist rim.

2. Combine the tequila, cider, and syrup in a shaker and fill with ice. Gently shake. Carefully pour into the prepared glass and garnish with the remaining lemon slice and the lavender sprig.

Lavender Simple Syrup
Makes 1 cup

- ½ cup water
- 2 tablespoons dried lavender blossoms
- 1 cup sugar

Bring the water and lavender to a boil in a saucepan. Add the sugar and stir until dissolved. Simmer for 5 minutes, then remove from the heat and allow to cool to room temperature. Strain out and discard the lavender, then bottle and refrigerate. The syrup will keep in the refrigerator for up to 2 weeks.

INGREDIENTS

- 2 lemon slices, reserving 1 for garnish
  Sugar, for garnish
- 1½ ounces tequila
- 3 ounces 101 Cider House Black Dog cider
- ½ ounce Lavender Simple Syrup
  Lavender sprig, for garnish

THE CIDER

BLACK DOG

This is the world's first "black cider." A title not yet seen outside these cider house walls, the cidery uses local Ventura County lemons with activated charcoal (a derivative of coconut) to blend the darkness with a touch of light into the apple base.

Maple Basil Ciderita

Recipe by Sonoma Cider, Healdsburg, CA

David Cordtz knows a thing or two about the craft beverage industry — he's started and sold a tea brand as well as a nonalcoholic sparkling cider company, and he was the winemaker at a winery he owned with his brother. Today, he is the CEO, founder, and cidermaker, along with his son Robert, for Sonoma Cider and sits on the board for the United States Association of Cider Makers. The family business is based in the heart of Northern California's apple country, allowing the father-son partnership to indulge in the area's agricultural riches, with access to West Coast organic fruit. As a "cider version of a Mexican Bulldog" (a margarita with a full, upside-down beer garnishing the glass), this cider-spiked margarita will satisfy any craving for cocktails south of the border.

DIRECTIONS

Makes 1 cocktail

Add the basil leaves, tequila, lime juice, and maple syrup to a pint glass and hand press with a muddler. Fill a highball glass with ice, strain in the muddled ingredients, and top with the cider. Garnish with the basil sprig.

INGREDIENTS

- 5 basil leaves
- 1 ounce silver tequila
- ½ ounce lime juice
- ½ ounce light maple syrup
- 5 ounces Sonoma Cider The Hatchet cider
- Basil sprig, for garnish

THE CIDER

THE HATCHET

The cidery ferments juice from organic apples for this cider, resulting in an off-dry sipper that reveals purity in apple fruit and a smooth yet crisp palate. A clean addition to the cocktail, this cider is drawn to the citrus, savory sweetness of the basil, and the sugar of the maple syrup, all while soothing the pang of the tequila.

Caledon Fizz

Recipe by David Greig, the Black Hoof and Grey Gardens, Toronto, ON

"I wanted this drink to showcase the cider at its best," says David Greig, cocktail programmer extraordinaire for five bustling bars in Toronto and one in Montréal, including the new cider-centric Grey Gardens. The decorated bartender colors outside the lines of a basic margarita, calling his libation a "simple homage to the provenance" of the featured cidery, Spirit Tree, based in the township of Caledon, Ontario. "The cider turns the drink from a regular margarita variant into something more refreshing and seasonally appropriate, while also adding the complexity you'd expect from a fermented beverage."

DIRECTIONS

Makes 1 cocktail

Pour the cider, tequila, lime juice, grapefruit juice, and honey syrup into a collins glass. Add ice and stir lightly. Garnish with the grapefruit wedge.

INGREDIENTS

- 2 ounces Spirit Tree Estate Cidery Draught cider
- 1 ounce blanco tequila
- ¾ ounce lime juice
- 2 teaspoons grapefruit juice
- 2 teaspoons honey syrup (2 parts honey, 1 part water)
- Grapefruit wedge, for garnish

THE CIDER

DRAUGHT CIDER

from Spirit Tree Estate Cidery, Caledon, ON
Founded on and named for the spiritual rituals of an English wassail — a cider-dredged ceremony that chases evil spirits out of the orchard in the dead of winter — Spirit Tree's Draught Cider is a medium-dry tipple comprising dessert and heirloom apples from the area. Crisp and straight with apple fruit, the cider is a complementary base for this citrus-soaked tequila drink.

Juan's Manzana

Recipe by Lindolfo Silva, The Painted Burro, Somerville, MA

Wordplay, spice, herbs, and fruit come together in this recipe donning a Spanish name for Johnny Appleseed, created by barman Lindolfo Silva of The Painted Burro. The Mexican cantina wafts fragrant and flavorsome scents through its open, street-side floor-to-ceiling windows, offering authentic Mexican cuisine and drink to the Boston suburb. Lightly oaked tequila meets its match with rosemary-and-chipotle-infused syrup and a float of Boston-based cider in this cocktail, providing both heat and refreshment.

Makes 1 cocktail

1. Line the rim of a rocks glass with a cinnamon-sugar mixture: Place the mixture on a small plate, rub the rim of the glass with a wet paper towel, and then place the rim in the mixture. The mixture will adhere to the moist rim.

2. Combine the tequila, syrup, bitters, and cider in a shaker and fill with ice. Shake and strain into the prepared glass, top with fresh ice, and garnish with the apple slice.

Rosemary-Chipotle Syrup
Makes about 3½ cups

1 ounce fresh rosemary (1–2 sprigs)
5 dried chipotle chilis
2 cups water
2 cups sugar

Combine the rosemary, chilis, water, and sugar in a saucepan. Bring to a boil, stirring until the sugar dissolves, and let simmer for 10 minutes. Strain before using.

Cinnamon-sugar mixture
1½ ounces reposado tequila
1 ounce Rosemary-Chipotle Syrup
4 dashes plum bitters
2½ ounces Bantam Cider Wunderkind cider
Apple slice, for garnish

WUNDERKIND
from Bantam Cider Company, Somerville, MA
Just across town is Bantam Cider, the city-based cidery with a mission to break standards and embrace the modern movement of the beverage. The signature cider is the Wunderkind, named for pioneer Amelia Earhart and featuring a blend of New England apples. Slow-fermented with a time-tested strain of yeast and splashed with flower-blossom honey, Silva finds that the crisp cider "brings the drink alive with the bubbles."

Rio to Rollinsford

Recipe by Erin Mahoney, Row 34 Portsmouth, Portsmouth, NH

Inspired by the product made by the restaurant's cider neighbor to the north, North Country Hard Cider, Row 34 Portsmouth's bar manager Erin Mahoney says the two come from a similar mind-set of farm-to-table, or bottle, and appreciates the like-minded local foraging the cidery does for its fruit. With the powerful and smoky addition of mezcal, the acid of the lemon juice, and the sweetness of the cider, apple brandy and cinnamon syrup help the cocktail find its footing somewhere between smoky and sweet, somewhere between Rio and Rollinsford.

DIRECTIONS

Makes 1 cocktail

Combine the mezcal, brandy, lemon juice, syrup, and bitters in a shaker. Strain into a Champagne flute. Top with the cider and garnish with the lemon twist.

Cinnamon Simple Syrup

Makes 2–3 cups

- 8 grams cinnamon bark
- 2 cups water
- 2 cups sugar

Wrap the cinnamon in a cloth and break it up with a muddler, then weigh it on a scale to be sure you have 8 grams. Lightly toast the cinnamon in a medium-size saucepan over medium-low heat. Add the water and sugar, stir, and heat to 180°F (80°C). Allow to reduce by about a third. Remove from the heat, cool, cover, and refrigerate for 24 hours. Strain before use.

INGREDIENTS

- ¾ ounce mezcal
- ¾ ounce Laird & Company AppleJack brandy
- ½ ounce lemon juice
- ½ ounce Cinnamon Simple Syrup
- 2 dashes Angostura bitters
- 1 ounce North Country Hard Cider Original Press cider
- Lemon twist, for garnish

THE CIDER

ORIGINAL PRESS

from North Country Hard Cider, Rollinsford, NH
This young cidery comprises three home cidermakers turned professionals, producing an all-natural cider of simply local juice and yeast on their New England cider mill. The Original Press, the cidery's flagship cider, is a blend of six varieties and pours crisp, clean, and refreshing as the topper to this quaffer.

Hard Tellin'

Recipe by Jeremy Walker, Il Villaggio Osteria, Teton Village, WY

Jeremy Walker might find himself a long way from his Vermont origin, but with the wide distribution of Woodchuck Hard Cider, he is able to get a taste of home at his Italian slopeside restaurant in Jackson Hole's Teton Village. Marrying the sweet Caribbean syrup of falernum with the piquant powers of fresh ginger juice, the caramel tones of the dark rum, and the baked apple of the amber cider all served in a pint glass, it's hard tellin' the potency this potable packs.

 DIRECTIONS

Makes 1 cocktail

Pour the cider, rum, falernum, ginger juice, lime juice, and bitters into a pint glass. Add ice and stir. Garnish with a lime wedge.

 INGREDIENTS

- 6 ounces Woodchuck Hard Cider Amber cider
- 2 ounces dark rum
- 1½ ounces falernum
- ½ ounce fresh ginger juice
- 1 ounce lime juice
- 3 dashes Angostura bitters
- Lime wedge, for garnish

THE CIDER

AMBER

from Woodchuck Hard Cider, Middlebury, VT
Amber started it all. Fermented in a two-car garage in 1991, Woodchuck's inaugural cider was the humble beginnings for one of the country's largest and most successful cideries to date. The amber cider still bottles flavors of semisweet, juicy red fruit that, with the rum and ginger, offers a spiced caramel apple profile perfect for fall.

The Force

Recipe by Jesse Card, Bit House Saloon, Portland, OR

A self-proclaimed *Star Wars* nerd, Jesse Card brought his geekdom to the glass in this collaboration of cider and rum. General manager of the Bit House Saloon, a Portland bar focused on single-barrel spirits, Card enjoys a selection of nearly 600 bottles of liquor at his disposal, pairing classic molasses-rich rum with local Oregon cider. "The cocktail is named 'the Force' because of the separation of the light and dark substances or 'sides,'" says Card. "And although the 'light side' of the Force is the more voluminous and sharp, the 'dark side' adds its rich flavor to every sip."

DIRECTIONS

Makes 1 cocktail

Measure the rum, falernum, and lime juice into a small measuring cup and stir to combine. Pour the chilled cider into a rocks glass. Slowly float the rum mixture onto the top of the cider. Squeeze the orange peel over the drink to express the oils, then discard. Garnish the drink with a light dusting of nutmeg.

INGREDIENTS

- 1 ounce Cruzan Black Strap rum
- ¼ ounce falernum
- ¼ ounce lime juice
- 6 ounces (¾ cup) Carlton Cyderworks Citizen cider, well chilled
- Wide strip of orange peel
- Freshly grated nutmeg, for garnish

THE CIDER

CITIZEN

from Carlton Cyderworks, McMinnville, OR

This family-owned cidery walks a fine line between traditionally inspired and modern ciders, even using the ancestral British spelling of *cyder*. The Citizen, an English-style semi-dry cider, presses more than a dozen English cider apples, like Kingston Black, Dabinett, and Yarlington Mill, fermenting into a rich and moderately tannic sipper that clings to the rum like long-lost friends.

NORTH AMERICAN CLASSICS

CIDER AND BROWN LIQUOR

Ask any cider-wise mixologist and they will tell you the most straightforward spirit counterpart is another North American paradigm — whiskey or bourbon. Maybe it's the agricultural connection of barley, wheat, corn, or rye with apple, or maybe it's just historical nostalgia, but the matching seems matchless in these brown liquor and cider cocktails.

The Whiskey Smash

Recipe by Corey Polyoka, Woodberry Kitchen, Baltimore, MD

When Corey Polyoka started at Woodberry Kitchen, the farm-focused eatery located in an old, brick-clad factory in Baltimore, he was behind the bar and brought the restaurant's strong relationships with local farms back there with him. Tweaking drinks to "reflect the fruits" of Woodberry's sourcing, this cocktail centers on local wildflower honey, verjuice (pressed juice of unripened grapes) from nearby wineries, and a house cider made by Distillery Lane Ciderworks.

 DIRECTIONS

Makes 1 cocktail

In a tall glass, drop in the bitters onto ice and top with the cider. In a shaker, combine the bourbon, verjuice, and syrup. Shake for 10 seconds, then strain twice before adding to the glass. Bruise the rosemary to release the oils, and set it in the glass as garnish.

INGREDIENTS

- 2 dashes barrel-aged bitters
- 1½ ounces Distillery Lane Ciderworks Woodberry cider
- 1¾ ounces Smooth Ambler Spirits Old Scout straight bourbon whiskey
- ¾ ounce verjuice
- ½ ounce wildflower honey syrup (8 parts honey, 2 parts water)
  Rosemary sprig, for garnish

THE CIDER

WOODBERRY CIDER
from Distillery Lane Ciderworks, Jefferson, MD
Spanning 95 acres in the historic town of Jefferson, Distillery Lane's farm dates back to the Civil War, when it was used as a stopover for Union soldiers. Years later during Prohibition, the property and area was a hotbed for moonshining, and today, it makes history for housing the state's first licensed cidery with more than 3,000 trees planted for cider. Production includes a house offering it makes for Woodberry, infused with aronia berries that grow wild throughout the acreage.

Stone Fence

Recipe by Graham Hogan, Maria's Packaged Goods & Community Bar, Chicago, IL

Liquor store and cocktail bar hybrid Maria's has been a multigenerational watering hole since 1986. Codirector of cocktails, spirits, and wine Graham Hogan commemorates the bar's history with a cocktail chronicle in this revamped, lighter version of the classic Stone Fence with Virtue, a Chicago cider favorite. "There is a lot of resonance between the ingredients that make for a really pleasant drinking experience," Hogan says. Don't be intimidated by the drink's foam: the four ingredients are simply loaded into a whipped cream dispenser that can be purchased from Amazon or any major kitchen supply store.

DIRECTIONS

Makes 1 cocktail

Combine the cider, ginger beer, and tea concentrate in a shaker and stir gently for about 30 seconds. Pipe a layer of foam into the bottom of a chilled coupe glass, and gently pour the cocktail through the center. The foam will float to the top. Garnish with the rosemary and candied ginger.

Bourbon-Maple Foam
Makes about 1½ cups

- 7 ounces egg whites
- 2 ounces dark maple syrup, chilled
- 6 ounces bourbon, chilled
- Pinch of salt

Combine the egg whites, maple syrup, bourbon, and salt in a whipped cream dispenser and charge with two nitrous oxide cartridges while shaking vigorously. The foam has a better structure when the ingredients are cold, so be sure the items have been chilled in the fridge, or refrigerate the blended ingredients in the canister for an hour before serving. Shake again before using.

INGREDIENTS

- 4 ounces Virtue Cider Mitten cider
- 1 ounce ginger beer
- 1 ounce FINE Brand Korean Honey Ginger Tea concentrate
- Bourbon-Maple Foam
- Rosemary sprig and candied ginger, for garnish

THE CIDER

MITTEN
from Virtue Cider, Fennville, MI

Greg Hall has an affinity for cider cocktails. Founder of Virtue Cider and former brewmaster of superstar Goose Island, Hall knows his way around the beverage world, and his barrel-aged ciders are sought-after acquisitions for cider geeks across the country. A "winter" cider, the bourbon barrel–aged Mitten was a requirement for this snow-season cocktail, a finalist in Virtue's 2015 Stone Fence Competition in Chicago.

Apple Cart

Recipe by Lauren Mote, Bittered Sling Bitters and UVA Wine & Cocktail Bar, Vancouver, BC

"When discussing all things apple, it's important to me to not only use apple-related products but complementary flavors that taste incredible alongside apples," says Lauren Mote, the coproprietor of bitters and tincture producer Bittered Sling and bar manager at the ultrasleek UVA in Vancouver. For the celebrated Canadian cocktail queen, the Apple Cart is a liquid embodiment of this stance on true flavors. "It is a combination of great ingredients, topped with a funky, tart, and refreshing cider," she says of the cocktail. "Bourbon has an inherent baked apple note, sherry's got a bit of crisp, fresh apple on the nose . . . and spiced apple on the palate." For Mote's tannic syrup, dried hops can be found at a homebrew store or online.

Makes 1 cocktail

Combine the bourbon, Aperol, sherry, lemon juice, syrup, and bitters in a shaker with ice. Strain into a tall collins glass filled with fresh ice. Top with the cider, stir gently, and serve garnished with a bushy bouquet of lemon verbena and mint.

Tannic Apple Syrup
Makes 2 cups

- 1½ cups honey
- 1½ cups apple juice
- ⅛ teaspoon dried hops

In a saucepan, bring the honey and apple juice to a low simmer. Remove from the heat and pour over the hops in a heatproof container. Allow to steep and infuse for 1 hour at room temperature. Strain and discard the hops, and bottle the syrup. It will keep, refrigerated, for 7 to 10 days.

⬦▶ **INGREDIENTS**

- 1½ ounces Bulleit bourbon
- ½ ounce Aperol
- ½ ounce Lustau East India Solera sherry
- ¾ ounce lemon juice
- ¾ ounce Tannic Apple Syrup
- 2 dashes Bittered Sling Zingiber crabapple bitters
- 4 ounces Tree Brewing Dukes Cider Dry Apple cider
  Bouquet of lemon verbena and mint, for garnish

⬦▶ **THE CIDER**

DUKES CIDER DRY APPLE
from Tree Brewing, Kelowna, BC

Cider made from a brewery in the center of British Columbia's wine country, Dukes Dry Apple is produced exclusively from apples grown in Kelowna — now a hotbed for cider fruit. Dry, tart, and lightly carbonated, this cider does the trick for Mote's multifaceted sipper.

Pear-fect Rye Fizz

Recipe by Kathy Casey, Kathy Casey Liquid Kitchen, Seattle, WA

Nationally renowned bar-chef and resident CIDERCRAFT cocktail columnist Kathy Casey found a mutual partnership in perry and rye in this classic whiskey fizz. "Reminiscent of classic poached pear flavors, the brown sugar and vanilla syrup add rich notes to complement the lighter flavors of the perry," she says. "Vander Mill's Bon Chrétien is a dry, tart perry with light bubbles and adds a fizz to a modern play on a whiskey sour."

DIRECTIONS

Makes 1 cocktail

Measure the rye, syrup, lemon juice, and bitters into a cocktail shaker. Fill the shaker with ice, cap, and shake vigorously. Strain the cocktail into a large coupe and top with the perry. Twist the lemon peel over the drink to release the oils, then discard. Garnish with the pear slice.

Brown Sugar–Vanilla Syrup
Makes 12 ounces

- 1 cup brown sugar, firmly packed
- 1 cup water
- 1 vanilla bean

Combine the sugar and water in a small saucepan. Split the vanilla bean pod down the center. Scrape the soft center of the vanilla bean out with a paring knife. Add the split pod and the soft vanilla bean paste to the pan and stir. Bring to a boil over medium-high heat, then quickly remove from the heat. Let cool to room temperature. Store refrigerated with the vanilla bean in the syrup.

INGREDIENTS

- 1½ ounces rye whiskey
- ¾ ounce Brown Sugar–Vanilla Syrup
- ¾ ounce lemon juice
- 2 dashes Peychaud's bitters
- 2 ounces Vander Mill Bon Chrétien perry
  Fat piece of lemon peel
  Thin slice of fresh pear, for garnish

THE CIDER

BON CHRÉTIEN
from Vander Mill, Spring Lake, MI

A founding producer to the recent revival of Michigan cider, Vander Mill began as a modest cider mill in 2006 and has since surged into markets in multiple states, with a second production and taproom across the state in Grand Rapids. The only perry in the cidery's diverse anthology, Bon Chrétien is an homage to the original name of the common Bartlett pear, "Good Christian."

Turnbull Tipple

Recipe by Sarah Ellis, Sycamore Den, San Diego, CA

An addition to San Diego's killer cocktail arena, Sycamore Den opened in 2013 and earned its reputation for putting a contemporary spin on classic cocktails. Garnering inspiration from the late 1970s "middle-class living room," the walls sport shotguns (that are nonfunctioning) and a blazing fireplace (that is functioning), while the cocktail menu highlights America throughout the decades, including light beer of the '70s and ciders of today. Former general manager Sarah Ellis takes this archetypal blend of scotch and something carbonated and puts a cider slant on it, melding the cider's sweetness with the spice of the ginger and smoke of the whiskey.

DIRECTIONS

Makes 1 cocktail

Combine the scotch, lemon juice, syrup, and sliced strawberry in a shaker and fill with ice. Shake, then strain into a highball glass with fresh ice. Top with the cider and garnish with a strawberry and lemon wheel.

Ginger Syrup

Makes 1½ cups

- 1 cup sugar
- 1 cup water
- 5 ounces fresh ginger, peeled and roughly chopped

Bring the sugar and water to a boil in a saucepan, stirring until the sugar is full dissolved. Add the ginger, bring to a simmer, then remove from the heat to steep for 1 hour. Strain and bottle the syrup. It will keep, refrigerated, for up to 30 days.

INGREDIENTS

- 1 ounce blended scotch
- ½ ounce lemon juice
- ½ ounce Ginger Syrup
- 2 strawberries, 1 sliced, 1 for garnish
- 2 ounces Tilted Shed Graviva! Semidry cider
  Lemon wheel, for garnish

THE CIDER

GRAVIVA! SEMIDRY
from Tilted Shed Ciderworks, Windsor, CA
Husband-and-wife duo Scott Heath and Ellen Cavalli are actively trying to bring the apple back to enocentric Sonoma County, California, with heirloom and cider apples planted on their wine country estate. The Graviva! is half Gravenstein and half heirloom and cider apples, a combination that results in bright acid, easy tannin, and an affable effervescence ideal for this libation.

Cardamomagin

Recipe by Andrew Volk, Portland Hunt + Alpine Club, Portland, ME

Well-traveled bartender Andrew Volk brought his cocktail and cider knowledge from the East Coast to the West and back again, returning to Portland, Maine, after time spent behind the bar in Oregon's Portland. At the Portland Hunt + Alpine Club in Maine, his praised bar program profiles flavors of the region, like cider, and beyond, like scotch. "Apples, scotch, and cardamom all play well together in an apple-cobbler sort of way," he says of his cocktail, spotlighting the "warming spices that make you want to snuggle by a fireplace in fall."

DIRECTIONS

Makes 1 cocktail

Combine the scotch, lemon juice, and syrup in a tall collins glass. Fill with ice and top with the cider. Garnish with lemon peel.

Hunt + Alpine Coriander-Cardamom Syrup
Makes 1½ cups

- 1 cinnamon stick, broken up
- 2 tablespoons coriander
- 2 tablespoons black cardamom
- ¼ teaspoon red-pepper flakes
- ½ cup Bacardi 151 rum
- ½ cup plus 2 tablespoons hot water
- ½ cup plus 2 tablespoons sugar

Place the cinnamon, coriander, cardamom, pepper flakes, and rum in a blender. Blend on low to medium speed for 2 to 4 minutes. Meanwhile, mix together the water and sugar to form a simple syrup. Strain the spice mixture into the syrup, mix, and refrigerate until ready to use.

INGREDIENTS

- 1½ ounces scotch
- ½ ounce lemon juice
- ¼ ounce Hunt + Alpine Coriander–Cardamom Syrup
- 4 ounces Urban Farm Fermentory Dry Cidah cider
- Lemon peel, for garnish

THE CIDER

DRY CIDAH
from Urban Farm Fermentory, Portland, ME
Enthusiasts of all things fermentable, Urban Farm Fermentory brings its love of yeasts-in-action to cider in its downtown Portland "experimental fermentation center." Fresh-pressed Maine apples undergo spontaneous fermentation to complete dryness, making this cider a true expression of the local "culture." For Volk, the dry cider adds depth and length to the spiced cocktail, broadening the flavors of the scotch in an approachable M.O.

CIDER, FROM START TO FINISH

APERITIFS, BRANDIES, AND LIQUEURS

Lower-alcohol refreshments have gained popularity, possessing all the flavor in an aperitif or liqueur fashion with a fraction of the booze content. This allows for cider to be an amiable bedfellow for a beverage union, highlighting its bolder traits. On the flip-side, empowered by apple-based distillates, cider and brandy also make nice on a higher ABV playing field.

Oh, Snap

Recipe by Meghan Garry, Citizen Cider, Burlington, VT

When Meghan Garry first started at Citizen Cider's tasting room in Burlington's South End Art District, cocktails were a routine of bourbon and cider. "We set out to create a cocktail list that incorporated one of our ciders into every single drink, which was fun but pretty challenging," Garry admits. "The inspiration really came from trying to think outside of the cocktail box and experiment with different flavors." Cue the ginger liqueur and Aperol — the Italian aperitif comprising bitter orange, gentian, rhubarb, cinchona, and more — combined with cider to create what Garry calls "a slight bitter edge and gingersnappy kick."

DIRECTIONS

Makes 1 cocktail

Combine the Aperol, ginger liqueur, lemon juice, and apple juice in a shaker with ice. Shake well. Strain into a highball glass with fresh ice. Top with the cider.

INGREDIENTS

- ½ ounce Aperol
- ¼ ounce Art in the Age (Ginger) SNAP liqueur
- ¼ ounce lemon juice
  Splash of apple juice
- 3 ounces Citizen Cider Unified Press cider

THE CIDER

UNIFIED PRESS

A year-round staple for this modern cidery, Unified Press is Citizen's "cider for the people." Only locally sourced fruit goes into this cider, a crisp, refreshing, and effervescent tipple that brings juicy apple fruit to the snap of the ginger and tang of the Aperol in a fresh, simple, and straight-forward cocktail.

Basil Bellini

Recipe by Eden Specialty Ciders, Newport, VT

The bone-chilling frigidity of northern Vermont in winter is an anticipated curse for residents but a blessing for the resident apples. Mirroring the ideal conditions of Québec's prevailing ice cider industry, Eden Specialty Ciders was the first to produce the handcrafted beverage in the United States. Ice cider led to other bottlings, like the Orleans Aperitif Ciders, "farm-to-bar alternatives to the more industrial European aperitif wine" used in cocktails. In this cocktail, a classic Bellini gets an herbal twist by way of Vermont cider.

DIRECTIONS

Makes 1 cocktail

Combine the honey and hot water in a shaker until smooth. Muddle the basil leaves in the honey mixture, add ice, and add the cider. Stir well then strain into a Champagne flute. Top off with the Prosecco and garnish with the basil sprig.

INGREDIENTS

- ½ teaspoon honey
- ½ teaspoon hot water
- 2 large basil leaves
- 2 ounces Orleans Herbal Aperitif Cider
- 1 ounce Prosecco
  Basil sprig, for garnish

THE CIDER

ORLEANS HERBAL APERITIF

Free of dyes, artificial sweeteners or flavorings, the Orleans aperitifs are pure cider and herb/root infusions, true expressions of the land from fruit to root. The Herbal takes concentrated yet dry cider and infuses it with a proprietary blend of Vermont-grown herbs, resulting in a fresh and aromatic sipper that brings floral, honey, and anise tones to the Bellini.

Pear and Blackberry Sangria

Recipe by Kathy Casey, Kathy Casey Liquid Kitchen, Seattle, WA

"Think sangria, and typically wine comes to mind," says Kathy Casey, leader of the bar-chef movement, founder of Kathy Casey Liquid Kitchen in Seattle, and CIDERCRAFT magazine's cocktail mastermind. "Think cider, and apple comes to mind. So, in this nontraditional sangria, I featured Crispin's Lion Belge. Its lovely and complex flavor profile is delicious made into a sangria highlighted with the flavors of honey and cognac." Note that the berry fruit in the recipe is interchangeable — such as raspberries instead of blackberries — depending on the season and local market.

DIRECTIONS

Makes 4–6 servings

1. Combine the water, lemon juice, and blackberries in a blender. Process until smooth. Strain through a fine-mesh strainer and discard the solids. Add the honey, orange juice, and brandy, and stir until the honey is dissolved. Refrigerate for up to 2 days.

2. When ready to serve, combine the blackberry mixture with the chilled cider in a large container. Serve over ice, garnished with fresh berries and pear slices.

INGREDIENTS

- ¼ cup water
- 2 tablespoons lemon juice
- ½ cup blackberries
- 1½–2 tablespoons blackberry honey
- 2 tablespoons fresh orange juice
- 2 tablespoons brandy or cognac
- 1 (22-ounce) bottle Crispin Cider Natural Hard Pear Cider Lion Belge, chilled
  Fresh blackberries and sliced pear, for garnish

THE CIDER

PEAR LION BELGE
from Crispin Cider Co., Colfax, CA

A household name for cider, Crispin's diversions from their main-line ciders come fast and often, like the limited release Lion Belge, where fresh-pressed, unfiltered pear juice is fermented with a witbier (wheat beer) yeast strain amid orange peel and coriander. Spiced, citrusy with fresh pear, this pear cider falls in line with the summer flavors of sangria.

Pippin's Stone

Recipe by Shawn Soole, cocktail consultant and mixologist, Victoria, BC

Shawn Soole's name is synonymous with cocktails in British Columbia and is connected with mixology across Canada. The lionized Australian bartender and cocktail consultant calls Victoria home and has cemented his handle and cachet in the capital city's cocktail heritage since he arrived. When it comes to bubbles in his cocktails, Soole says he prefers sparkling cider instead of soda. "It gives you the texture you need and adds even more flavor," he explains. Pippin's Stone, spritzed with cider and spiked with apple brandy from two nearby cideries, takes a hit of concentrated apple with the body of lemon juice for a refreshing and spicy sipper.

DIRECTIONS

Makes 1 cocktail

Combine the wine, brandy, syrup, and lemon juice in a shaker with ice. Shake well. Strain and pour into a highball glass with fresh ice. Top with the cider. Garnish with the lemon twist.

INGREDIENTS

- 1½ ounces Stone's Original Green Ginger Wine
- ½ ounce Merridale Ciderworks apple cider brandy
- ½ ounce simple syrup
- ¾ ounce lemon juice
- 3 ounces Sea Cider Farm and Ciderhouse Pippins cider
  Long lemon twist, for garnish

THE CIDER

PIPPINS

from Sea Cider Farm and Ciderhouse, Saanichton, BC
A proponent of agritourism, Vancouver Island's Sea Cider brings the people to the cider — a 10-acre cider apple farm and cider house Soole is personally familiar with. A "sharp" cider due to the Yellow Albemarle (Newton) Pippins and Champagne yeast used to ferment them, this robust sipper stands up to the spice of the ginger and concentrated strength of the local apple cider brandy.

Dugspur Try

Recipe by Matthew Beason, Black Twig Cider House, Durham, NC

Quaffable, crushable, and guilt-free: Matthew Beason's low-alcohol cider refresher can be drained from the glass at his cider-centric Durham bar, Black Twig. Named for Andrew Jackson's favorite variety, the bar version of the Tennessee apple came to North Carolina in 2016 by way of Beason, cider enthusiast and restaurateur. The cocktail mixes Foggy Ridge Cider's dessert cider with aromatic wine-based aperitif Byrrh and a Durham-made cucumber shrub to revivify Southern cider sipping.

 DIRECTIONS

Makes 1 cocktail

Combine the cider, Byrrh, and shrub in a mixing glass. Add ice and the soda water to a rocks glass, then pour in the cider mixture. Top with the bitters.

 INGREDIENTS

- 1 ounce Foggy Ridge Cider Pippin Gold cider
- ½ ounce Byrrh aperitif
- ½ ounce Crude cucumber shrub
- 2 ounces soda water
- 3 drops of Crude Bitterless Marriage bitters

THE CIDER

PIPPIN GOLD

from Foggy Ridge Cider, Dugspur, VA

Cider pioneer and trail-blazer for female makers across the globe, Diane Flynt's Foggy Ridge ciders have been at the pinnacle of orchard-based ciders in the United States. As a dessert rendition of the popular Albemarle (Newtown) Pippin apple, the Pippin Gold is no different, blending apple brandy from the country's oldest distillery and Virginia local, Laird & Company. In this libation, its higher proof lends concentrated sweetness without tipping the scale on this lighter, lower alcohol drink.

American 75

Recipe by Tim Buzinski, Artisan Wine Shop, Beacon, NY

Food and beverage industry vet Tim Buzinski goes all-American, all the way, in his take on a French 75. "Once I tasted the Aaron Burr Ginger cider, I knew it would make a great component in a cocktail," Buzinski says of finding his muse for revisions on this classic drink. "The applejack sets the apple tone for the drink, while the lemon and simple syrup bring some nuance. The cider makes the drink with its gingery imprint and the subtle apple tones it lends in support of the applejack."

DIRECTIONS

Makes 1 cocktail

Combine the applejack, lemon juice, and syrup in a shaker with ice. Shake well and strain into a collins glass filled with fresh ice. Pour the cider over the top and gently stir to combine. Garnish with an apple slice.

INGREDIENTS

- 2½ ounces Harvest Spirits Cornelius applejack
- ½ ounce lemon juice
- 1 ounce simple syrup
- 5 ounces Aaron Burr Cider Ginger cider
- Apple slice, for garnish

THE CIDER

GINGER

from Aaron Burr Cidery, Wurtsboro, NY

Remember Aaron Burr? The New Americana vice president who fatally shot former secretary of the treasury Alexander Hamilton in a famous duel in New Jersey? Owner and cidermaker Andy Brennan certainly does, enough to name his cidery after the fermented-apple drinker himself. Aspiring to produce New World cider reminiscent of those times, the Aaron Burr Ginger is a fusion of the past and present, fermenting New York apples with fresh Chinatown ginger for a piquant yet rustic cider.

Strawberry Fields

Recipe by Ambrosia Borowski, The Northman, Chicago, IL

It's hard not to hum the Beatles' psychedelic hit of a similar name when sipping this cider from Chicago's first cider bar, The Northman. Ambrosia Borowski transitions the melody into the glass with a medley of summer fruit and cross-country flavors. "My inspiration for this cocktail is the memory of a fresh strawberry lemonade being sipped slowly on a hot summer day," she notes. With apple brandy from a local distillery, Borowski channels the sunshine with strawberry cider from nearby Wisconsin.

DIRECTIONS

Makes 1 cocktail

Combine the brandy, cider, lemon juice, and syrup in a mixing glass with ice. Strain and serve over fresh ice in an old-fashioned glass. Garnish with the thyme sprig.

INGREDIENTS

- 2 ounces Rhine Hall Distillery apple brandy
- 3 ounces Ciderboys Hard Cider Strawberry Magic cider
- ½ ounce lemon juice
- ½ ounce simple syrup
  Thyme sprig for garnish

THE CIDER

STRAWBERRY MAGIC
from Ciderboys Hard Cider, Stevens Point, WI
Answering the Midwestern call for more local cider, the duo behind Wisconsin's Ciderboys bottles "fruit-ful pairings" of apple and alternate flavors, like cinnamon bark and pineapple, for the masses. A splash of strawberry allows this cider to embrace the fruits of summer and juiciness needed for the simple yet bright cocktail.

The Fruity Pot of Tea

Recipe by Fabien Maillard, Le Lab Comptoir à Cocktails,
Montréal, QC

Leading barman Fabien Maillard knows the finer etiquette
of libations, from tea to cocktails, and showcases good
form through his vintage cocktail bar, Le Lab, with
flair-bartending and handcrafted tipples. Maillard finds
inspiration for this cider cocktail in an early seventeenth-
century punch consumed by Colombian nobility. "It's a
wintry mix that is comforting and rich with exotic and
sunny aromas," Maillard details. "The aroma of vanilla
that is found in maple syrup combines well with apple and
mango, and the ice cider adds a touch of roundness and
acidity."

DIRECTIONS

Makes 1 cocktail

Combine the maple syrup, lemon juice, mango, Drambuie,
vanilla liqueur, and cider in a shaker with ice. Shake and
strain twice into a teacup or glass. Garnish with a dehydrated
apple slice.

INGREDIENTS

- ¾ ounce maple syrup
- 1 ounce lemon juice
- 1 ounce mango purée
- ¾ ounce Drambuie
- ¾ ounce Galliano vanilla liqueur
- ¾ ounce Cidrerie CRYO Mi-CRYO ice cider
  Dehydrated apple slice, for garnish

THE CIDER

MI-CRYO
*from Cidrerie CRYO,
Mont-St-Hilaire, QC*
Located at the base of
Mont-Saint-Hilaire, in the
heart of the Montérégie
region, CRYO was founded
in 2007 at the source of
ice cider's origin. The
Mi-CRYO is a blend of
McIntosh, Spartan, Golden
Russet, and Empire apples
that undergo cryoconcen-
tration (when the juice
from late-harvest dessert
apples is pressed in winter
and allowed to freeze) for a
still, caramel-rich ice cider
that adds the necessary
fuller body to this "tea"
with a nip of acid.

TABLESIDE SIPPERS

RECIPES FOR COOKING AND PAIRING WITH CIDER

"Farm-to-table" has new meaning when it comes to pairing with cider, and an even more substantial significance when cider is used in the actual cooking. North America's top chefs took on the challenge of cider-based cuisine, detailing, mapping out, and pairing so you can do the same in your kitchen at home. Ingredient availability might vary by your location, but the adaptable chefs offer suggestions for substitutions when needed.

FIRST COURSE

APPETIZERS, SALADS, AND SIDES

Kimchi Fried Rice Balls

Recipe by Chi'Lantro, Austin, TX

Champions of kimchi, Austin's Chi'Lantro is "Korean barbecue inspired," fusing Mexican and Texas influences with the pickled, the spicy, and the flavorsome. By way of nearly half a dozen brick-and-mortar shops and a food truck, Chi'Lantro brings its eclectic, multicontinental cuisine to the people of the live music capital. These kimchi-spiked fried rice balls use gochujang, a traditional Korean chili paste, which can be found at most Asian specialty markets.

DIRECTIONS

Makes 24 balls

1. In a skillet set over medium heat, combine the kimchi, cooked rice, gochujang, cheese, and toasted sesame oil. Sauté, tossing to mix everything together, until the cheese has melted, about 7 minutes. Spread on a baking sheet to cool. Once cool enough to handle, roll into balls, using about 2 tablespoons per ball.

2. Whisk the egg in a shallow bowl. Arrange the flour and breadcrumbs in separate shallow bowls. Dip each ball into the flour, then the beaten egg, then the breadcrumbs.

3. In a deep pot over medium heat, heat the oil to 350°F (180°C). Fry the balls in small batches, turning until golden brown on all sides, about 5 minutes. Drain on paper towels and serve.

THE PAIRING

The Texas-born team behind Texas Keeper Cider brings the flavors of the Lone Star State to the bottle, naming the ranch-based cidery after a native apple. The Ciderweizen merges the green fruits and citrus flavors of the apples with the spicy, floral hoppiness of a German hefeweizen beer. The multifaceted cider complements the many flavors of the snack, providing crispness to the crunch of the rice ball.

INGREDIENTS

- 1 cup kimchi, finely chopped
- 2¾ cups cooked sushi rice
- 1 tablespoon gochujang
- 2 cups white cheddar cheese, shredded
- 1½ teaspoons toasted sesame oil
- 1 jumbo egg, beaten
- ⅔ cup all-purpose flour
- ⅔ cup panko breadcrumbs
- 3–4 cups vegetable oil, for frying

Virginia Apple Mostarda

Recipe by Chef Joe Sparatta, Heritage, Richmond, VA

Raised in the kitchen by culinary megastars like Eric Ripert and David Bouley, chef Joe Sparatta has received accolades of his own since starting his restaurant in Richmond. Often collaborating with esteemed cidermaker Diane Flynt of nearby Foggy Ridge Cider, he features her cider with a range of flavors and Virginia apples to compile a regional *mostarda* — a classic Italian "condiment" using fresh fruit and mustard. Sparatta recommends serving the *mostarda* over pork loin or simply on a crostini with a mild, semifirm cow's milk cheese, like the Appalachian from Meadow Creek Dairy in Galax, Virginia.

DIRECTIONS

Makes 1½ to 2 cups

1. In a large saucepan, combine the vinegar, cider, sugar, ginger, salt, and pepper over high heat and bring to a boil.

2. Add the shallots and mustard seeds. Decrease the heat to medium and cook until the mixture is reduced by half, 10 to 15 minutes. Add the apples and cook until the mixture becomes syrupy, about 20 minutes.

3. Remove from the heat, add the mustard oil, and stir to combine. Serve warm or at room temperature.

INGREDIENTS

- 8 ounces apple cider vinegar
- 1½ ounces Foggy Ridge Cider Serious Cider
- ½ plus ⅛ cup sugar
- 1½ ounces fresh ginger, grated
- 1 teaspoon sea salt
- ½ teaspoon freshly ground black pepper
- 6 shallots, cut into thin rings
- 3 tablespoons yellow mustard seeds
- 2 pounds Virginia Stayman Winesap apples (or another tart apple variety), peeled and finely diced
- 1–2 drops mustard oil

THE PAIRING

One of the original Virginia cideries in the modern era, Foggy Ridge was founded in 1997 by pioneer Diane Flynt. Growing old Virginia apples like Hewes Crab and Harrison with bitter cider apples like Ashmead's Kernel and Dabinett, Foggy Ridge's Serious Cider blends the best of both worlds into a crisp and light tipple that complements the cooked apple, tangy mustard seeds, and zesty ginger of the *mostarda*.

Smoked Salmon and Asparagus Hush Puppies
with Saffron-Cider Aioli

Recipe by Chef/Owner Gabriel Schuenemann, Alder Wood Bistro, Sequim, WA

Sheltered in the rain shadow of the Olympic Mountains, "Sunny Sequim" is an intimate bayside, densely agricultural town in Washington's Olympic Peninsula. Chef Gabriel Schuenemann's charming downtown bistro caters to the farmer and the foodie, with wood-fired plates and regional, organic, and sustainable cuisine. In his original recipe, Schuenemann uses saffron from Phocas Farms in nearby Port Angeles and freshly ground corn from Sequim's Nash's Organic Produce but suggests organic store-bought brands if not shopping in the Olympic Peninsula.

DIRECTIONS

Makes 36–40 hush puppies

1. Prepare the aioli: In a bowl, crumble the saffron and steep in the cider for 30 minutes, then whisk in the aioli.

2. Prepare the hush puppies: Fill a small electric fryer or cast-iron skillet about a third of the way up with cooking oil, and heat to 375°F (190°C).

3. While the oil heats, whisk together the cornmeal, flour, chili powder, baking powder, baking soda, salt, and pepper in a large bowl. In a separate bowl, whisk together the egg, buttermilk, honey, carrot, asparagus, and onion. Pour the egg mixture into the flour mixture and stir until just mixed, then fold in the salmon. The batter should be thick like cake batter, not runny like pancake batter. If it is too thick, add a little water.

INGREDIENTS

SAFFRON-CIDER AIOLI

Pinch of saffron (about 15 stamens)

2 tablespoons Finnriver Farm & Cidery Solstice Saffron cider, or fresh organic lemon juice

1 cup aioli or mayonnaise

HUSH PUPPIES

Canola oil

1 cup freshly ground cornmeal

1 cup all-purpose flour

1 teaspoon Piment d'Espelette chili powder, or cayenne

1 teaspoon baking powder

1 teaspoon baking soda

1 teaspoon sea salt

Freshly ground pepper

1 egg

1 cup buttermilk

1 tablespoon honey

1 small carrot, diced

5 asparagus spears, trimmed and sliced into rounds

½ red onion, diced

5 ounces smoked salmon, flaked or roughly chopped

Additional salt and pepper, for seasoning

4. To form the hush puppies, use a small ice cream scoop or two spoons, filling one with the batter and using the other to scrape it into the hot oil. Fry, turning occasionally, until golden, 2 to 3 minutes. Remove with a slotted spoon and season immediately with salt and pepper. Break open samples from the first batch to make sure the fritters are cooked through, and adjust the cooking time as needed.

Farm focused and hyper-local, Finnriver Farm & Cidery uses the same source of saffron for its Solstice Saffron cider. The semisweet blend of organic heirloom and dessert apples from Washington is delicately steeped in the rare Mediterranean threads for a robust and warmly spiced drink. Not shy on acid, the cider lights up the rich fritters and makes for a scrumptious multicultural snack.

Frisée Salad with Pickled Strawberries

Recipe by Chef Kenji Hurlburt, The Queens Kickshaw, Queens, NY

People pack into the sardine-size, narrow gastropub that is The Queens Kickshaw, a counter service–styled eatery that plates embellished comfort classics. Broadcasting respected beer and cider lists next to searing espresso, chef Kenji Hurlburt offers a cider vinegar–laced frisée salad recipe perfect for summer. Note the strawberries require a minimum of 24 hours to pickle.

DIRECTIONS

Makes 4 servings

1. Prepare the pickled strawberries: Combine the water, vinegar, lemon juice, sugar, star anise, and ginger in a small pot over medium heat and bring to a boil. Once the sugar dissolves, remove from the heat and let cool. In a bowl, pour the cooled liquid over the strawberries; refrigerate for at least 24 hours. Reserve ½ cup of pickling liquid.

2. Prepare the vinaigrette: In a small bowl, whisk together the pickling liquid, honey, Dijon, and salt. Slowly whisk the oil into the mixture until it emulsifies.

3. Assemble the salad: In a large bowl, mix the greens and nuts with 1 cup of the prepared pickled strawberries and ½ cup of the strawberry vinaigrette. Season with salt to taste. Top with the feta and drizzle with the balsamic reduction.

THE PAIRING

Fellow New Yorker Original Sin Hard Cider launched small in 1997 and is now in over 30 states and counting. Dry and effervescent, the flagship Original Sin cider offers invigorating fruit flavors to the bright and tangy pickled strawberries, while the earth of the apple matches the greens and the acid cuts through the feta.

INGREDIENTS

PICKLED STRAWBERRIES

- 1 cup water
- 1 cup apple cider vinegar
- 1½ teaspoons lemon juice
- ½ cup sugar
- 2 pieces star anise
- 1 small knob fresh ginger, peeled and cut in half
- 1 pint strawberries, tops removed, cut in quarters

STRAWBERRY VINAIGRETTE

- ½ cup reserved pickled strawberry liquid
- 1 tablespoon honey
- 1 tablespoon Dijon mustard
- 1 teaspoon salt
- 1 cup grapeseed or canola oil

SALAD

- 4 cups frisée or other mixed young greens
- ½ cup toasted walnuts
- Salt
- ¼ cup crumbled feta cheese
- 2 tablespoons balsamic reduction

Charred Cabbage Salad

Recipe by Mei Mei Street Kitchen, Boston, MA

Run by a trio of siblings — *mei mei* means "little sister" in Chinese — this creative Chinese-American street food restaurant and food truck showcases the family ties with the bonding of unique flavors. Dedicated to serving food out of "unexpected places," Mei Mei also repurposed a shipping container into an avant-garde walk-up window for their innovative grub. The charred cabbage in this salad brings a smoky complexity to a basic produce item, while the apples, feta cheese, egg, and garlic-toasted breadcrumbs offer a myriad of textures and tastes.

DIRECTIONS

Makes 4 servings

1. In a small bowl, whisk together the vinegar and chili oil to taste. Slowly pour in the olive oil while whisking. Set the dressing aside while you make the salad.

2. Preheat a grill to high heat. Meanwhile, add canola oil to a skillet set over medium heat. Stir in the garlic. Add the breadcrumbs and sauté until golden brown. Season with a pinch of salt and set aside.

3. Place the cabbage halves cut side down on the grill. Char for 2 to 3 minutes. Remove from the heat and slice into thin ribbons. Transfer the cabbage and apples to a large bowl; toss with a generous amount of the dressing. Add the breadcrumbs and feta; season with salt to taste. Divide the salad between four plates.

4. Melt the butter in a large skillet set over medium-low heat. Crack the eggs directly into the pan and cook until the whites are set and the yolks are still runny, 2 to 3 minutes. Top each serving of salad with an egg and serve.

INGREDIENTS

- ¼ cup rice wine vinegar
- Chili oil
- ¾ cup olive oil
- 2 tablespoons canola oil
- 1 garlic clove, minced
- ½ cup coarse panko breadcrumbs
- Kosher salt
- 1 small green cabbage, halved
- 2 red apples, cored and diced
- ½ cup crumbled feta cheese
- 1 tablespoon butter
- 4 eggs

THE PAIRING

Downeast Cider House is a definitive New England cidery, and its Original Blend cider lays claim as a throwback to the farm-based ciders that once prospered in the area. Fresh with a clean apple flavor up front, the hints of sweetness in both salad and cider meld together for a seasoned fusion of New England tastes.

Shaved Brussels Sprouts Salad

with Smoked Trout, Black Walnuts, Pecorino, and Hard Cider Vinaigrette

Recipe by Chef Ben Poremba, Elaia, St. Louis, MO

Restaurateur and chef Ben Poremba's St. Louis food empire stretches throughout the Gateway City, including Elaia, the intimate, fine-dining restaurant adjacent to his casual, vinyl-record-spinning wine bar Olio. Rooted in Mediterranean influence with seasonality of his home region, Poremba's dish is basic at the crux but is enhanced with flavor combinations like smoky fish with sharp cheese and thinly shaved Brussels sprouts, a childhood nightmare vegetable turned cool.

⬧ DIRECTIONS

Makes 4–6 servings

1. Preheat the oven to 350°F (180°C). Spread the walnuts on a baking sheet and bake for 8 to 10 minutes.

2. In a large bowl, combine the Brussels sprouts, chives, trout, pecorino, and walnuts. Lightly toss with the vinaigrette, and serve.

Hard Cider Vinaigrette
Makes about 1 cup

  2 cups Urban Chestnut Brewing Bushelhead cider
  ¼ cup apple cider vinegar
  ¼ roasted walnut oil
  ⅓ cup grapeseed oil
  1 teaspoon Dijon mustard

In a saucepan over medium-low heat, cook the cider until it reduces to ⅓ cup, roughly the consistency of maple syrup. In a bowl, whisk together the cider, vinegar, walnut oil, grapeseed oil, and mustard. Store in the refrigerator.

⬧ INGREDIENTS

  ½ cup black walnut pieces
  4 cups Brussels sprouts, thinly shaved on a mandoline
  1 cup chives, thinly sliced
  ½ cup smoked trout, picked over and flaked
  ½ cup pecorino, in chunks
  ¼–½ cup Hard Cider Vinaigrette

⬧ THE PAIRING

A hometown St. Louis brewery goes rogue with its first cider exposé. Urban Chestnut Brewing Company's Bushelhead is easy drinking and approachable, a golden goblet sold at a brewery and made by a brewer to help band the two separate bodies of cider and beer. Reducing the cider burns off a bit of the booze in this recipe but keeps the intense and fermented flavors, adding complexity to the vinaigrette and making a natural partner in pairing.

Randolph's Polenta

Recipe by Randolph's Restaurant & Bar, Warwick Denver Hotel, Denver, CO

A hidden gem in Denver's uptown, Randolph's in the Warwick provides patrons with a variety of plates from regional mountain cuisine to English pub influences. For the namesake polenta, a broad contrast of flavors is put together, from bitter winter greens to the sweet roasted grapes and buttery polenta.

DIRECTIONS

Makes 6 servings

1. Preheat the oven to 350°F (180°C). Line a baking sheet with parchment paper.

2. Blanch the walnuts in a pot of boiling water for 8 minutes. Drain but do not dry. While still wet and warm, toss them in a bowl with the sugars to coat, then bake for 10 to 12 minutes.

3. Bring the 6 cups of water to a boil in a large saucepan and stir in the salt. Gradually whisk in the cornmeal. Reduce the heat to low and cook until mixture thickens and the cornmeal is tender, stirring often, about 15 minutes. Remove from the heat. Add the butter and stir until melted. Cover and keep warm.

4. In a large skillet, heat 3 tablespoons of the oil over medium heat. Add the garlic and onion and stir for 2 minutes. Add the kale and spinach, season with salt and pepper to taste, and sweat the greens for 5 minutes.

5. Preheat the broiler. Wash and dry the grapes. Gently combine the grapes, the remaining 1 tablespoon oil, and salt and pepper, to taste, in a bowl. Spread on a baking sheet and broil for 3 to 5 minutes, until lightly charred.

6. To serve, divide the warm polenta between six bowls. Top with a generous portion of the kale mixture, then a handful of roasted grapes and candied walnuts. Finish each serving with plenty of Gorgonzola crumbles.

INGREDIENTS

- 1 cup shelled walnuts
- ¼ cup granulated sugar
- ¼ cup brown sugar
- 6 cups water
- 2 teaspoons salt
- 1¾ cups yellow cornmeal
- 3 tablespoons unsalted butter
- 4 tablespoons olive oil
- 8 garlic cloves, finely minced
- ½ white onion, finely chopped
- 1 bunch chopped kale
- 1 6-ounce bag baby spinach
  Salt and pepper
- 2 cups red seedless grapes
- 1½ cups crumbled Gorgonzola

THE PAIRING

Warm, creamy, and roasty flavors seek similar support in Colorado Cider Company's Glider Cider, a sipper that exhibits fresh-sliced apple aromas, baking spices, and an easy tartness from the Colorado-grown apples. The flagship for the cidery, this cider's approachable flavors match the candied walnuts and grapes, while the acidity lasers through the rich polenta.

Cider-Glazed Carrots

Recipe by Poverty Lane Orchards and Farnum Hill Ciders, Lebanon, NH

Apples have been on the Poverty Lane Orchards' fields since the 1960s, with cider fruit planted in the 1980s — some of the oldest cider apple trees in the country. This recipe for simple pan-cooked carrot spears tells the story of the land much like the orchard's Farnum Hill ciders aspire to. Glazed in the cidery's fruit-forward semi-dry cider, earthy and terroir-driven flavors come from the drink and the root vegetable.

DIRECTIONS

Makes 4 servings

In a large pan, combine the carrots, cider, water, butter, maple syrup, vinegar, thyme, cayenne, and salt and black pepper to taste. Bring to a simmer over medium-low heat and cook until most of the liquid evaporates and the carrots are crisp-tender, 25 to 30 minutes. Remove the carrots from pan if they finish cooking before the liquid is reduced. Remove and discard the thyme. Adjust the seasoning and serve the carrots with the reduced liquid as a sauce.

THE PAIRING

In cooking with the Farnum Hill Semi-Dry, the cider is an obvious pairing for the carrots glazed in its juices. Golden in hue and flavors, the tropical fruits and citrus rind mimic the same tastes in the dish as they do the glass. The least dry offering from Farnum Hill, this cider's full body and ample astringency balance the sweetness of the carrots and match the flavors it lends to the plate.

INGREDIENTS

- 1 pound carrots (about 6), peeled and cut into 2-inch batons
- 1 cup Farnum Hill Cider Semi-Dry cider
- ¼ cup water
- 2 tablespoons unsalted butter
- 1 tablespoon maple syrup
- 2 teaspoons cider vinegar
  Thyme sprig
  Dash of cayenne pepper
  Salt and freshly ground black pepper

Wild Mushroom Spaetzle

Recipe by Chef/Owner Dan Cellucci, Roots Cafe, West Chester, PA

Spaetzle, the soft egg noodle dumpling, has many names and iterations depending on its origin — from Austria and Switzerland to German-influenced Pennsylvania. Owner Dan Cellucci saw a void for a seasonal restaurant in his town of West Chester, and he opened Roots Cafe with the mission to plate sustainable, seasonal, and local food to the community savvy to its regional Pennsylvania Dutch cuisine. Spaetzle follows suit with wild mushrooms — shiitakes, chanterelles, or morels will all offer a savory earthiness — and a healthy dose of chèvre and herbs from the rolling hills of the region.

DIRECTIONS

Makes 4 servings

1. Combine the flour, salt, pepper, thyme, and garlic powder in large bowl. Whisk together the eggs and milk in a separate bowl. Slowly pour the milk mixture in a steady stream into the flour mixture, stirring to form a wet and sticky dough.

2. Bring a large pot of water to boil over high heat. Pour dough into a colander. Using a spatula, press the dough through the colander holes so the pieces fall directly into the boiling water. After 2 minutes, remove the spaetzle and rinse under cold water to stop the cooking.

3. In a 10-inch skillet, heat the butter and oil over medium-high heat. Add the mushrooms, garlic, and shallot, and sauté for 2 minutes. Stir in the arugula and spaetzle, and season with salt and pepper to taste. Sauté for 3 minutes, then shake the pan to flip the spaetzle as they turn golden brown. Add the chèvre and stir gently. Serve immediately.

INGREDIENTS

- 3 cups flour
- 1 tablespoon salt
- 1 tablespoon cracked black pepper
- 1 tablespoon minced fresh thyme leaves
- 1 tablespoon garlic powder
- 6 eggs
- ¾ cup whole milk
- 1 tablespoon butter
- 1 tablespoon olive oil
- 2 cups diced wild mushrooms or shiitakes
- 1 garlic clove, minced
- 1 shallot, finely diced
- 1 cup chopped arugula
- Salt and freshly ground black pepper
- ¼ cup chèvre

THE PAIRING

Just north of historic Gettysburg and a short distance from the Appalachian Trail, Big Hill Ciderworks is amid orchards in the hub of Pennsylvania's Apple Country. Its heirloom-based cider production is showcased in the Standard, a bright and apple-forward cider with acid to counter the creaminess of the chèvre and earthy fruit to match the mushrooms.

Grilled Pimento Cheese Sandwiches

Recipe adapted from Mercier Orchards, Blue Ridge, GA

Started by Bill and Adele Mercier in 1943, Mercier Orchards is a fourth-generation family-owned and -operated farm, cidery, and deli in Georgia's historic High Country. In the kitchen, the team whips up lunch favorites like this grilled pimento cheese sandwich, oozing with Southern grit. The pimentos lend a tangy flavor to the cheddar, affirming the combination's title as the "caviar of the South," while the corn relish adds another dimension of farm-fresh sweetness. If the orchard bakery isn't a short drive away, grab simple country bread to spread the cheese on and serve with an apple slaw.

INGREDIENTS

- 2 cups shredded cheddar cheese
- ½ cup good-quality mayonnaise
- 1 (4-ounce) jar chopped pimentos, drained
- ¼ cup prepared corn relish
- 12 slices thick white bread
- 2 tablespoons butter, melted

DIRECTIONS

Makes 6 servings

1. In a food processor, pulse to combine the cheddar, mayonnaise, pimentos, and corn relish. Chill for at least 3 hours.

2. Divide the pimento cheese among six slices of bread and spread into a thick layer. Top with the remaining bread. Brush the bread lightly with the butter.

3. Preheat a skillet over medium heat. Grill the sandwiches (as many at a time that fit comfortably) until the bread is toasted and the cheese is soft, about 3 minutes per side. Serve warm.

THE PAIRING

Founder Adele Mercier's spirit lives on in the cider that carries her name, Adele's Choice. Bubbly and dry, the apples for this cider are the first to go to the press, resulting in a crisp and slightly tart sipper. That astringency provides a necessary sharpness to pierce the unctuousness of the gooey, rich sandwich.

Grilled Asparagus and Radicchio
with Hazelnuts and Lemon Cream

Recipe by Chef Justin Large, Vander Mill, Spring Lake and Grand Rapids, MI

Tenured Chicago chef Justin Large has spent time in the kitchens of several James Beard–nominated restaurants, has appeared on the Food Network's *Iron Chef* and Travel Channel's *No Reservations,* and is adorned with media nods from around the country. His culinary skills are on display at Vander Mill cidery's two in-house restaurants. In this charred veg-heavy salad, he highlights the power of cider and fresh produce, without the use of dairy in the "cream."

INGREDIENTS

- 2 heads Treviso radicchio, cleaned and halved
- 4 tablespoons extra-virgin olive oil
- 1 bunch asparagus, trimmed
- 1 cup coarsely crushed hazelnuts
- ¼ cup Hazelnut Vinaigrette
  Salt and freshly ground black pepper
- ¼ cup Lemon Cream
- 1 tablespoon lemon zest

DIRECTIONS

1. Preheat a grill. Toss the radicchio in 2 tablespoons of the oil and char on the grill rack cut side down. Cool and cut into ½-inch-wide ribbons. Toss the asparagus spears in the remaining 2 tablespoons oil, and char on the grill to your preference of doneness, 5 to 7 minutes.

2. In a dry skillet, toast the hazelnuts until browned, 3 to 5 minutes.

3. In a large mixing bowl, toss the radicchio, asparagus, and half of the nuts together. Add the vinaigrette, season to taste with salt and pepper, and toss to incorporate. Plate the salad on a large serving platter. Drizzle with lemon cream and top with the remaining hazelnuts and the lemon zest.

THE PAIRING

Vander Mill has been working the food and cider circuit since 2006. For a smoky but bright salad, the light but richly flavored Vander Mill Hard Apple corroborates the vigor of the cider both in the "cream" and in the can. Easy acidity matches that of the lemon and enhances the char of the vegetables without fighting the green flavors of the asparagus.

Hazelnut Vinaigrette

Makes 1 cup

- 1 teaspoon chopped fresh thyme
  Salt and freshly ground black pepper
- 1 medium shallot, diced
- ¼ cup Champagne vinegar or white wine vinegar
- ¾ cup hazelnut or olive oil

1. Sprinkle the thyme and salt and pepper to taste over the shallot, and mince well. Transfer to a medium bowl, and stir in the vinegar. Macerate for 5 minutes.

2. While mixing with a whisk, slowly drizzle the hazelnut oil into the shallot mixture. Incorporate well, then taste and adjust for seasoning and balance.

Lemon Cream

Makes 1 cup

- 1 egg yolk
- 2 tablespoons Vander Mill Hard Apple cider
- 4 cloves roasted garlic
  Juice of 2 lemons
  Salt and freshly ground black pepper
- 1 cup canola oil

In a food processor or blender, combine the egg yolk, cider, garlic, half of the lemon juice, and salt. With the processor or blender running, slowly drizzle in the oil until well incorporated. The mixture should be homogeneous but not thick. Loosen the mixture with the remaining lemon juice and with water, if necessary; season to taste with salt and pepper.

FROM THE WATER
FISH, MOLLUSKS, BIVALVES, AND MORE

Grilled Mackerel
with Napa Cabbage

Recipe by Chef/Owner Earl Ninsom, Paa Dee and Langbaan, Portland, OR

Known for his definitive Thai cuisine accentuated with Pacific Northwest flair and ingredients, Bangkok native Akkapong "Earl" Ninsom goes at his Portland Thai food two ways — comforting and finger-licking classics by Paa Dee and exotically orchestrated tasting menus at Langbaan. Not to mention that guests enter the latter by walking through a trick bookcase door in the back of the former. Ninsom's restaurants are unique and ingredient driven, much like this simple and sauced recipe that speaks Thai through Portland fare.

Makes 4 servings

1. In a medium saucepan over medium heat, combine the water, soy sauce, sugar, and rice wine. Stir until the sugar dissolves, then bring to a boil for about 1 minute. Cover and remove from the heat.

2. Preheat a skillet over high heat and add 1 tablespoon of the oil. Generously season the fillets with salt and pepper. Place on the skillet skin side down for 3 minutes, then flip to finish cooking. When done, the fish will be opaque white and gently flake apart when pressed with a knife blade. Remove from the heat.

3. In a wok set over high heat, stir-fry the garlic, cabbage, and the remaining 2 tablespoons oil until smoking hot and slightly charred on the edges. Divide the cabbage mixture between four plates and top each plate with two mackerel fillets and the sweetened soy sauce. Garnish with the scallions and cilantro.

INGREDIENTS

- 2 cups water
- 1 cup soy sauce
- ¾ cup sugar
- ¼ cup mirin rice wine
- 3 tablespoons soybean oil
- 4 mackerels, skin on, filleted and deboned
  Salt and freshly ground pepper
- 1 teaspoon minced garlic
- 4 cups coarsely chopped napa cabbage
- 2 scallions, sliced, for garnish
  Fresh cilantro, for garnish

THE PAIRING

Reverend Nat's Deliverance Ginger Tonic takes a "futuristic" approach to the simple integration of apple and ginger. Each batch of this cider receives gallons of ginger juice, hundreds of hand-cut lemongrass stalks, juice and zest from thousands of limes, and hand-extracted quinine from Peruvian cinchona tree bark. The resulting tonic is zesty and revitalizing, juicy with a zip of spice and astringency, which sees that of the stir-fry sauces and tangy char of the cabbage.

Clams and Chorizo
with Sweet Peas and Leeks

Recipe by Chef Paul Zerkel, Goodkind, Milwaukee, WI

Tucked in from the water in Milwaukee's Bay View neighborhood, Goodkind's warm and welcoming beach house atmosphere makes it a cozy cavern to duck into after a day on the lake. With a menu deep in seafood (and rotisserie), chef Paul Zerkel also dives into the local gems of fermented fruit, steaming Manila clams and Spanish chorizo in a bath of semi-dry sparkling cider and accenting the dish with sweet peas and spring leeks.

DIRECTIONS

Makes 4 servings

1. Preheat the broiler. Set a large pot over medium heat and add the butter. Add the leeks, chorizo, and half of the parsley. Sauté until the leeks are soft and the chorizo is a little crispy, 4 to 5 minutes.

2. Increase the heat to medium-high. Add the clams and stir gently, until they are well coated. Add the cider and season with salt and pepper, then reduce the heat to medium-low. Cover the pot and let simmer until the clams open, about 5 minutes. Add the peas during the last minute.

3. While the clams are steaming, place the baguette slices under the broiler and toast until browned, about 2 minutes per side. Pour the clams and broth into a shallow serving bowl. Garnish with the remaining parsley, add the toasts, and serve hot.

INGREDIENTS

- 2 tablespoons unsalted butter
- 1 leek, diced and well rinsed
- 2 ounces Spanish chorizo, thinly sliced
- 1 bunch Italian parsley, minced
- 3 pounds Manila clams, rinsed, scrubbed, and soaked in salt water for 1 hour
- 1 cup ÆppelTreow Winery & Distillery Apply Doux cider Salt and freshly ground black pepper
- ½ cup fresh sweet peas (frozen is optional)
- 1 baguette, sliced and buttered

THE PAIRING

"True to the apple," ÆppelTreow Winery & Distillery's name hints at their mantra, crafting artisan ciders, like the Apply Doux, a semi-dry, sparkling by way of the Champagne method, and terroir driven by Red Delicious, Cortland, and other old Wisconsin baking apples. Its texture and fruit character make it a match to take on the spice of the chorizo and brininess of the clams.

Oven-Roasted Salmon
with Neige Onion Jam, Sunny Eggs, and Buttered Toast

Recipe by Danny St-Pierre, La Petite Maison, Montréal, QC

You might recognize Danny St-Pierre from television. Cohost of the traveling gourmand show *Ma Caravane au Canada* and host of the daily cooking program *Qu'est-ce qu'on mange pour souper?*, St-Pierre has become a household name for Canadian cuisine and is leading the "bistronomy" movement (elevated food in casual settings) in Montréal with his brunch-and-lunch-centered eatery, La Petite Maison. This oven-roasted salmon is lacquered with an ice cider–glazed onion jam, topped with an egg cooked sunny side up and positioned on crusty artisan bread. This is how "toast" should be.

DIRECTIONS

Makes 4 servings

1. Preheat the broiler.

2. In a skillet over medium-low heat, sweat the onions with 1 teaspoon of the oil and a pinch of salt until translucent, about 5 minutes. Deglaze the onion from the pan with the cider and cook until the liquid is reduced to a glaze, roughly another 5 minutes.

3. Rub the fish with the remaining 1 teaspoon oil and season with salt and pepper. Line up the fillets on a baking sheet, with the thyme placed under the flesh and the skin side up. Broil about 5 minutes, until the skin is charred. The flesh should be firm but with a moist center.

4. In a skillet, cook the eggs sunny side up. Toast the bread, then spread on the butter while the toast is hot. Spread glazed onion jam onto each salmon fillet and position on the bread. Top each toast with one egg. Garnish with scallions and a vigorous crack of black pepper.

INGREDIENTS

- 2 red onions, sliced
- 2 teaspoons vegetable oil
- Salt
- 1 cup Neige Première cider
- 4 (5-ounce) fillets fresh salmon, skin on
- Freshly cracked black pepper
- 4 sprigs thyme
- 4 eggs
- 4 slices country bread
- ½ cup (1 stick) salted butter
- ¼ cup finely chopped scallions

THE PAIRING

Domaine Neige is the Québec cidery that introduced ice cider to the world. Its Neige Première was the first ice cider to be labeled as such and still sets the standard today with flavors of yellow apples, maple, and lush peaches matched with brilliant acidity. Drawing those same qualities into the onion, the cider-glazed jam is a sunny and sweet addition to the brunch-time fish dish.

Nashi Orchards Perry-Steamed Mussels
with Fennel Confit, Pear Butter, and Fresh Pear

Recipe by Chef Sam Burkhart, Etta's, Seattle, WA

James Beard Award–winning restaurateur and chef Tom Douglas leads a distinguished team of culinary masterminds at his nearly two dozen establishments across Seattle, including Etta's, his eatery at the edge of historic Pike Place Market. Floor-to-ceiling windows gaze out onto the market, panoramic Puget Sound, and surrounding islands, serving as a creative influence for chef Sam Burkhart, who runs Etta's kitchen today. Burkhart recommends market-fresh crusty bread to soak up the delicious broth of these perry-steamed mussels and slather with unctuous confit.

FENNEL CONFIT

- 1 bulb fennel, cut into ¼-inch slices
- 2 cups olive oil
- 5 garlic cloves
- 5 sprigs thyme
- 2 bay leaves
- Zest of 1 lemon

MUSSELS

- ⅓ cup fennel confit (recipe above), roughly chopped
- 3 tablespoons reserved fennel confit oil
- ¼ cup thinly sliced onion
- 1 teaspoon sliced garlic
- 1 pound fresh mussels, washed and beards removed
- Pinch of salt
- 1 bottle Nashi Orchards Chojuro Blend perry
- 3 tablespoons pear or apple butter
- 3 tablespoons unsalted butter
- 1 teaspoon salt, more or less to taste
- 2 teaspoons lemon juice
- ¼ Asian pear, julienned fennel fronds, roughly chopped

DIRECTIONS

Makes 2 servings as a meal, 4 servings as an appetizer

1. Prepare the fennel confit: Combine the fennel, oil, garlic, thyme, bay leaves, and lemon zest in a small pot. The oil should barely cover the ingredients. Cook over medium-low heat so the oil is just bubbling and until the fennel is soft, about 45 minutes. Do not let the oil get too hot or it will fry the fennel instead of slowly cooking it. Let cool, then pick the fennel out of the oil and reserve for the mussels. Strain and discard the solids from the oil.

2. Prepare the mussels: Heat a skillet over high heat for 1 minute. Add the fennel confit, confit oil, onion, and garlic. Sauté until the onions start to develop color, 1 to 2 minutes. Add the mussels and a pinch of salt to the pan, toss, and cook 30 for seconds. Deglaze the pan with the perry, then cover the pan until the mussels start to open, 3 to 4 minutes.

3. Remove the lid and add the pear or apple butter and unsalted butter. Allow the liquid to reduce by a third, about 3 minutes. Add 1 teaspoon salt and the lemon juice to the reduced liquid. Stir, taste, and adjust the seasoning. Pour the mussels into a serving bowl and top with the julienned pear and fennel fronds.

THE PAIRING

One of the islands that gazes back at the Seattle waterfront is Vashon, home to Nashi Orchards and its nurtured plot of Asian pear trees. Only sold to a select number of restaurants off the island, Nashi's Chojuro Blend Asian pear perry is found at Etta's, and the soft, earthy flavors call for the complement of briny and meaty mussels.

Blackened Redfish and Pecan Romesco

Recipe by Wes Mickel, Argus Cidery, Austin, TX

Before he could be found in the cellar or in some of the country's oldest southern orchards, cidermaker Wes Mickel was in the kitchen. Working in restaurants around Austin and writing curricula for Whole Foods Market's culinary centers, Mickel's first love of food led him to cider. Combining his two passions, Mickel's recipe for blackened redfish is matched with a signature cider from his Austin cidery. This is "Spanish cider made with Texas ingredients, and Spanish food made with regional ingredients," Mickel says. If redfish is a hard find, snapper makes a great substitute. He suggests serving the fish with crusty bread to sop up the nutty romesco sauce.

DIRECTIONS

Makes 4 servings

1. Prepare the blackening seasoning: In a small bowl, combine the ancho powder, garlic powder, onion powder, oregano, and thyme. Store in an airtight container at room temperature.

2. Prepare the romesco: In a food processor or blender, combine the chilis, red pepper, garlic, vinegar, and salt. Purée until smooth, then add the pecans and bread. With the motor running, slowly incorporate the oil until the purée thickens.

3. Prepare the fish: Pat the fish dry and place on a clean plate. Season both sides with salt and pepper, then dust each fillet with roughly ½ teaspoon of the prepared blackening seasoning.

4. Preheat a skillet over medium heat with enough oil to lightly coat the bottom of the pan. Once the oil is warmed, add 2 seasoned fillets at a time. Once browned on the first side, flip the fish and add 1 tablespoon of the butter per fillet. When the second side is browned, remove the fish to a clean plate

INGREDIENTS

BLACKENING SEASONING

- ½ tablespoon ancho powder
- 1 teaspoon garlic powder
- 1 teaspoon onion powder
- ½ teaspoon dried oregano
- ½ teaspoon dried thyme

PECAN ROMESCO

- 3 pasilla chilis, toasted and deseeded
- 1 roasted red pepper, deseeded
- 6 garlic cloves, peeled
- 3 tablespoons apple cider vinegar
- ½ teaspoon salt
- ½ cup toasted pecans
- 1 slice bread, torn into small chunks
- ¼–½ cup extra-virgin olive oil

REDFISH

- 4 (8-ounce) redfish fillets
- Salt and freshly ground pepper
- Olive oil
- 4 tablespoons butter

and reserve the leftover oil to pour over the fish later. Wipe the pan clean and repeat the process with the remaining fillets.

5. To finish, serve each fillet on top of a heaping spoonful of romesco with a drizzle of the reserved cooking oil.

◈▶ **THE PAIRING**

Argus Cidery was one of the first producers on the map in Texas and was the first commercial maker to use Texas fruit. Sourcing from some of the South's most southern orchards, the Perennial is Texas fruit fermented naturally and aged in French and American oak barrels, resulting in a dry yet fruit-forward cider, influenced by the wild yeast flavors of Spanish *sidra* and matching the Catalan flavors of romesco.

Salmon Chowder
with Corn and Fennel

Recipe by Sharon Campbell and Olaiya Land, Tieton Cider Works, Yakima, WA

Tieton Cider Works founder and longtime orchardist Sharon Campbell is a firm believer in the power of pairings. When she teamed up with local chef Olaiya Land to create recipes specifically using her cidery's products, she found she had her favorites. One of her most satisfying collaborations is the duo's salmon chowder, a richly flavored broth, spiced and reddened with paprika and full of potatoes, corn, and salmon, matched with a juicy Yakima Valley cider.

DIRECTIONS

Makes 4–6 servings

1. In a large stockpot or Dutch oven, heat the oil over medium-high heat. Add the fennel seeds and cook, stirring, until fragrant, about 30 seconds. Add the onion, fennel, carrots, smoked paprika, and a generous pinch of salt, and cook, stirring frequently, for 1 minute. Reduce the heat to medium-low, cover, and cook until the vegetables are translucent and have softened a bit, about 5 minutes.

2. Add the cider and bay leaf and increase the heat to medium. Cook, uncovered, for 3 minutes. Add the milk, clam juice, potato, and corn. Bring the chowder to a bare simmer. Do not let it boil or it will break; it will taste fine but look curdled. Continue to cook at a simmer, stirring occasionally, until the vegetables are tender, about 30 minutes. Stir in the cream and black pepper, and bring the chowder back up to a low simmer.

INGREDIENTS

- 2 tablespoons olive oil, plus additional for garnish
- 1 teaspoon fennel seeds
- 1 small onion, diced
- ½ medium fennel bulb, cut into rough dice, fronds chopped and reserved for garnish
- 3 medium carrots, sliced into rounds
- ½ teaspoon smoked paprika
  Salt
- ½ cup Tieton Cider Works Wild Washington cider
- 1 bay leaf
- 2 cups whole milk
- 1 cup clam juice
- 1 large potato, peeled and cut into ½-inch dice
- 1½ cups corn (from about 2 cobs fresh corn or 1 [14-ounce] can)
- ½ cup cream
- ½ teaspoon freshly ground black pepper
- 1 pound wild salmon, skin and pin bones removed, cut into 1-inch pieces
- 2 tablespoons chopped parsley
  Ground cayenne pepper or paprika

3. Remove the chowder from the heat and add the salmon. Cover and let sit for 5 minutes off the heat. Serve the chowder with a drizzle of olive oil and a sprinkling of parsley, reserved fennel fronds, and cayenne.

THE PAIRING

Tieton Cider Works launched organically grown fruit from its beautifully situated Yakima Valley orchard in 2008, planting cider apple varieties that same year. The Wild Washington Apple is made with these varieties, originating from New England, England, and France, and producing a bittersweet cider angled at the American palate that can stand up to the creamy weight of this chowder.

Pan-Seared Halibut
with Cider Beurre Blanc and Spiced Delicata Squash

Recipe by Chef Daniel Kedan, Backyard, Forestville, CA

Sparked by the ranchers and farmers in its small Sonoma County town, Backyard promotes the whole-hog experience, buying locally raised meat and butchering in-house to serve with exclusively regional fare. So much so that the modest, hacienda-style restaurant touts a Michelin star for its hyperlocal, sustainable efforts on the dishes that are often plated on salvaged redwood. Chef Daniel Kedan tops his pan-seared halibut with a cider-soaked beurre blanc sauce — traditionally done with white wine — and spices up delicata squash with a Basque French dried pepper (feel free to substitute cayenne pepper, if needed).

DIRECTIONS

Makes 4 servings

1. Prepare the beurre blanc sauce: Combine cider, shallot, parsley stems, and peppercorns in a medium saucepan over medium-high heat. Bring to a boil and reduce the mixture until the cider is well concentrated and only a small amount of liquid remains. Reduce the heat to medium, add the cream, and cook until the mixture is reduced to just over a tablespoon.

2. Slowly whisk in the butter, a few pieces at a time. Stir in the lemon juice and season to taste with salt. Strain. Keep warm over very low heat.

3. Prepare the squash: Preheat the oven to 350°F (180°C). Line a baking sheet with parchment paper. Combine the paprika, Piment d'Espelette, pepper, and salt in a small dish. Place the squash pieces in a large bowl and toss with the oil. Sprinkle on the spice mixture and toss to coat. Place on the prepared baking sheet and roast for 10 to 12 minutes until tender and lightly browned.

Recipe continues on page 240

INGREDIENTS

BEURRE BLANC SAUCE
- 1½ cups Devoto Orchards Cider Save the Gravenstein cider
- 1 shallot, diced
- 6 parsley stems
- 1 tablespoon black peppercorns
- ¼ cup heavy cream
- 1 cup (2 sticks) butter, coarsely chopped
- 1 tablespoon lemon juice
- Salt

SQUASH
- ¼ cup sweet paprika
- ¼ cup Piment d'Espelette chili powder
- 1 tablespoon freshly ground black pepper
- 1 tablespoon salt
- 2 small Delicata squash, seeded and sliced into ¼-inch half-moons
- 2 tablespoons olive oil

HALIBUT
- 2 tablespoons grapeseed oil
- 4 (5-ounce) halibut fillets
- Salt

Pan-Seared Halibut, *continued*

4. Meanwhile, prepare the halibut: In a large ovenproof skillet, warm the oil over medium-high heat. Once hot, carefully place the fish fillets in the pan and sear until golden brown, about 1 minute. Transfer the skillet to the oven for 5 to 7 minutes, or until the halibut is cooked through and flakes gently under pressure. Season to taste with salt.

5. To assemble, divide the squash equally between four plates. Place a piece of halibut on top. Drizzle the perimeter of each plate with the beurre blanc.

THE PAIRING

Since the late 1970s, Devoto Orchards Cider has been a founding farmer to the regrowth of heirloom apples in Sonoma County, and the Save the Gravenstein honors that history of the eponymous apple. Farmed certified organic and fermented to dryness, this cider couples with the sweetness of the squash and richness of the halibut, but its bite of acid breaks through the opulence of the butter sauce and brings the dish together.

FROM THE LAND

BEEF, LAMB, PORK, AND POULTRY

Michigan Cassoulet

Recipe by Chef James Rigato, the Root Restaurant and Bar, White Lake, MI

A well-oiled, chef-powered machine, The Root Restaurant and Bar is seasonal cuisine prepared and plated to express Michigan terroir. Chef James Rigato knows this better than most — the Michigan native strives to bring his home state's food to the forefront, with hearty vegetable dishes and hearth-warming fare like this cassoulet. The slow-cooked French casserole recipe sticks to traditional plans of the dish, deviating slightly with goat confit and toppings of chèvre. Feel free to sub in the more commonly found duck confit, or try using shredded dark turkey meat.

DIRECTIONS

Makes 6 servings

1. Drain the beans and transfer to a large pot. Add the stock, salt pork, and herb sachet. Simmer gently, stirring occasionally, until beans are tender, about 45 minutes. Turn off the heat. Strain the beans, reserving the liquid.

2. Preheat the oven to 350°F (180°C).

3. In large pot over medium heat, melt the butter until foamy. Add the onion, celery, parsnip, and garlic, and sauté until translucent. Stir in the preserved lemon. Add the beans, along with enough reserved liquid to cover the beans and vegetables and look a bit soupy. Heat until almost boiling, then add the lemon juice and turn off the heat. This dish will be finished in the oven, where the beans will continue to absorb liquid. Do not discard the reserved bean liquid; you may need to add more during baking.

4. Place half of the bean mixture in a roasting pan. Arrange the confit and sausage evenly in a layer, and top with the remaining beans. Add more liquid if necessary to cover the beans. Bake, uncovered, for 30 to 45 minutes, until bubbly and thick but still fluid. Remove from the oven and sprinkle the

INGREDIENTS

- 4 cups dried white beans, soaked overnight in 1 gallon water
- 1 gallon chicken stock
- 2 ounces salt pork, pancetta, or bacon
  Sachet of thyme, parsley stems, and bay leaf
- ½ cup (1 stick) butter
- 1 cup diced onion
- 1 cup diced celery
- ½ cup diced parsnip
- 6 garlic cloves, minced
- 1 tablespoon preserved lemon, minced
  Juice of 1 lemon
- 1 pound goat or duck confit, cut into 4-ounce pieces
- 8 ounces Andouille sausage, sliced into ¼-inch coins
  Coarse breadcrumbs, toasted
- 2 tablespoons chopped parsley
- 2 ounces chèvre

breadcrumbs on top. Return to the oven and bake for 5 to 10 minutes. Sprinkle the finished dish with the parsley and chèvre just before serving.

THE PAIRING

Virtue Cider bottles earthy, rustic sippers inspired by European farm traditions. Using heirloom Michigan apples and barrel aging in French oak for nine months, the Percheron is named after the large, ancient breed of draft horse that historically worked the orchards of Normandy. Funky and round, this cider finds solace in food with a similar disposition, like this farmhouse stew.

Pork Meatballs
with Asparagus, Rhubarb, and Egg Sauce

Recipe by Butcher and Chef Collin Donnelly, LexMex Tacos, Lexington, VA

As a chef, butcher, and charcutier, Collin Donnelly has traveled extensively throughout the United States, coast to coast, learning what he can about regional cuisine along the way. His food path has taken him throughout the apple-growing regions and cider-centric states of the country, allowing him to fully grasp cooking with cider. This dish is a love letter to the culinary uses of cider, featuring it in every aspect except for the simply blanched asparagus.

DIRECTIONS

Makes 4 servings

1. Prepare the rhubarb: Place the rhubarb in a small mixing bowl. Combine the cider and sugar in a small saucepan and bring to a boil over high heat. Carefully pour the boiling cider mixture over the rhubarb. Let cool to room temperature, then strain the rhubarb from the cider. Set aside.

2. Prepare the egg sauce: Pour several inches of water into a saucepan and add the vinegar. Bring to a simmer over medium heat. Crack the eggs into a bowl and, one at a time, carefully drop each egg into the water. Cover the pan and turn off the heat. Let cook for 4 minutes. Gently transfer the eggs to a blender and add the mayonnaise, mustard, cider, lemon juice, and salt. Purée until smooth; refrigerate.

3. Prepare the meatballs: Put the bread in a shallow bowl and pour the milk over it. In a separate large bowl, combine the pork, pepper, Parmesan, thyme, and salt, tossing gently with your fingers. Add the soaked bread and cider to the meat mixture. Combine just until everything is evenly mixed.

INGREDIENTS

RHUBARB
- ½ pound rhubarb, cut on the diagonal into ¼-inch-thick slices
- ½ cup Castle Hill Cider Terrestrial cider
- 1 tablespoon sugar

EGG SAUCE
- 2 teaspoons white vinegar
- 3 large eggs
- 1 tablespoon mayonnaise
- 1 tablespoon yellow mustard
- 1½ teaspoons Castle Hill Cider Terrestrial cider
- 1 teaspoon lemon juice
- ½ teaspoon salt

MEATBALLS
- 2 slices wheat bread
- ⅓ cup milk
- 1 pound ground pork
- 2 teaspoons freshly ground black pepper
- 2 tablespoons grated Parmesan cheese
- 1 teaspoon chopped fresh thyme
- 1 teaspoon salt
- 2 tablespoons Castle Hill Cider Terrestrial cider
- 2 tablespoons olive oil

ASPARAGUS
- 1 bunch thick asparagus, trimmed
  Salt

4. With damp hands, form 12 small balls from the mixture. Add the oil to a large skillet set over medium heat. Cook the meatballs slowly, turning occasionally, until browned and cooked through, about 8 minutes.

5. As the meatballs cook, prepare the asparagus: Set a large pot of water over high heat and add a generous amount of salt. Bring to a boil. Add the asparagus and cook for 2 to 3 minutes. Drain promptly and rinse immediately with cold water.

6. To assemble, apply a swoosh of egg sauce to each plate, then arrange several stalks of asparagus and three meatballs on top. Add a few pieces of poached rhubarb and serve immediately.

THE PAIRING

A historic estate with ties to cider lovers Thomas Jefferson, George Washington, and James Madison, Castle Hill is also home to a cidery of the same name. The orchard grows old apples of prestige like Winesap and Albemarle (Newtown) Pippin, both featured in the Terrestrial, a crisp, bone-dry cider that cleanses the palate while slicing through these brawny meatballs and matching the tang of the rhubarb.

Goat Burgers
with Pickled Rhubarb Slaw

Recipe by Ian Gray, the Curious Goat, Minneapolis, MN

Found on four wheels and usually at Minneapolis's Sociable Cider Werks, chef Ian Gray mobilizes farm-to-fork cuisine with the Curious Goat food truck that focuses on caprine meat. He also brings in produce and more from local farmers' markets and features regional flavors with a daily fresh sheet marked on the chalkboard of the bright orange vehicle. This burly goat burger is jacked up with sweet and tangy pickled rhubarb and fresh slaw, placed between two buns, and smothered with chèvre.

DIRECTIONS

Makes 4 servings

1. Prepared the pickled rhubarb: Place the rhubarb in a medium bowl. Combine the water, vinegar, sugar, and salt in a small saucepan and bring to a boil over high heat. Carefully pour the hot brine over the rhubarb. Cool, cover, and keep refrigerated for up to 1 week. The rhubarb is best if prepared the day before serving. Strain the rhubarb when ready to use, reserving ½ cup of the pickling liquid.

2. Prepare the slaw: In a large bowl, gently combine the watercress, snow peas, radishes, scallions, pickled rhubarb, pickling liquid, and salt and pepper to taste. Set aside.

3. Prepare the burgers: Preheat the oven to 425°F (220°C). In a large bowl, mix together the meat, chèvre, chives, parsley, oregano, salt, black pepper, and pepper flakes. Form into four equal-size patties.

4. Heat an ovenproof skillet over medium-high heat. Add the oil. When hot, add the patties and cook for about 5 minutes. Flip the burgers and place the skillet in the oven for 3 minutes. Remove the pan from the oven, place the farmer's cheese

INGREDIENTS

PICKLED RHUBARB

- 2 cups thinly sliced rhubarb
- 1 cup water
- ½ cup rice wine vinegar
- ¼ cup sugar
- 1 tablespoon salt

SLAW

- ¼ pound watercress
- 1½ cups thinly sliced snow peas
- 1 cup sliced radishes
- 1 cup chopped scallions
- 1 cup pickled rhubarb
- ½ cup reserved pickling liquid
  Salt and freshly ground pepper

BURGERS

- 2 pounds ground goat
- 4 ounces chèvre
- ⅓ cup minced fresh chives
- ⅓ cup minced fresh parsley
- 1 tablespoon fresh oregano
- 2 tablespoons salt
- 1 tablespoon freshly ground black pepper
- 1 tablespoon red-pepper flakes
- ½ cup safflower oil
- 1 cup goat's milk farmer's cheese
- 4 buns

around the burgers, and return the skillet to the oven for 3 minutes, or until the burgers are cooked to your ideal temperature. Toast the buns while burgers are in the oven.

5. To finish, place one burger on the bottom of each bun. Stir the warm cheese in the skillet briefly on the stovetop over low heat, until melting. Top each burger with melted cheese and slaw, then with the remaining buns.

THE PAIRING

Fermenting a selection of Midwestern apple varieties, the innovative ciders from the urban Sociable Cider Werks vary from traditional to experimental — like the Spoke Wrench, a cider-stout hybrid that coferments fresh apple juice and stout brewer's wort with English ale yeast. Roasty and toasty with hits of juicy apple, the result is a worthy partner to the pickled burger combination.

Marinated and Grilled Flank Steak

Recipe by Chef Andrew Martin, Caffè Dolce at 500 Brooks, Missoula, MT

Missoula channels Italy in this sunny bistro with French doors that fling open to a cheerful patio and an equally bright menu. Featuring casual counter service by day and tableside service by night, Caffè Dolce offers chef Andrew Martin's menus of hand-cut pastas, antipasti, and Montana classics with an Italian spin, like this recipe for flank steak. Both the Italian-inspired *salsa verde* and marinated steak can get going the night before, but the radishes should be served hot with the finished steaks.

DIRECTIONS

Makes 4 servings

1. Prepare the steaks: In a large bowl or a baking pan, combine the garlic, oil, Worcestershire, and soy sauce. Coat the steaks with the mixture and marinate for at least 6 hours in the refrigerator.

2. Prepare the *salsa verde*: Roughly chop the parsley, cilantro, onion, garlic, and jalapeño then combine in a blender. Add the egg yolks, almonds, capers, olives, sherry, oil, lime juice, vinegar, and salt, and purée until thick and smooth. Cover and set aside.

3. Prepare the radish salad: Heat a large skillet over high heat, add the radishes, then toss with salt and pepper to taste. Sear until brown, about 5 minutes. Reduce the heat to medium and continue cooking, tossing often, until tender, about 5 minutes. Turn off the heat and let sit until the steaks are cooked and resting, then return to high heat. When the pan is sizzling, add the mustard greens and toss until wilted, 1 to 2 minutes.

INGREDIENTS

STEAKS

- 4 garlic cloves, minced
- ½ cup olive oil
- 2 tablespoons Worcestershire sauce
- 2 tablespoons soy sauce
  Kosher salt and freshly ground black pepper
- 4 (8-ounce) flank steaks

SALSA VERDE

- 1 bunch flat-leaf parsley, stems removed
- 1 bunch cilantro, stems removed
- ½ small white onion
- 3 garlic cloves
- 1 jalapeño, seeded
- 2 hard-boiled eggs, yolks only
- ¼ cup blanched almonds
- ¼ cup drained capers
- ¼ cup pitted green olives
- 1 cup dry sherry
- ½ cup olive oil
  Juice of 2 limes
- 1 tablespoon cider vinegar
- 1 teaspoon kosher salt

RADISHES

- 1 pound radishes, trimmed and quartered into wedges
  Kosher salt and freshly ground black pepper
- 1 quart fresh mustard greens or arugula, chopped

4. Preheat a grill to 475°F (245°C). Remove the steaks from the marinade and season heavily on both sides with salt and pepper. Cook the steaks to medium-rare, about 4 minutes per side. Rest for 5 minutes before slicing thinly across the grain. Serve with a generous amount of *salsa verde* and vegetables.

THE PAIRING

The North Fork Traditional serves as a flagship offering from Montana CiderWorks, the state's first cidery. A blend of hard-to-find traditional bittersweets and crab-apples, this semi-dry English-style cider highlights golden color and flavors. Faintly sweet with a smack of sharpness, the tart fruit mingles with the *salsa verde*, and the tannin complements the rich steak.

Cider-Braised Chicken

Recipe by Owner Annemarie Ahearn, Salt Water Farm Cafe and Market, Lincolnville, ME

Cooking school, garden, farm, and kitchen, Salt Water Farm Cafe and Market is led by the classically trained chef and owner Annemarie Ahearn. After spending time in Parisian kitchens and working under celebrity restaurateur Tom Colicchio, Ahearn moved to Maine to build a recreational cooking school, restaurant, and market. Simple and farm fresh, this whole roasted chicken is a union of flavors, influences, and cider Ahearn collected along the way.

DIRECTIONS

Makes 4–6 servings

1. Preheat the oven to 350°F (180°C). Blot the chicken with paper towels and season with salt and pepper. Place the flour in a shallow bowl. Dredge the chicken pieces in flour, one at a time, shaking off excess flour. Set them on a plate.

2. Place the bacon in a large cast-iron pan. Cook over medium-low heat until most of the fat is rendered and the bacon is crisp, 10 minutes. With a slotted spoon, transfer the bacon to a plate lined with paper towels.

3. Increase the heat to medium and add the butter to the pan. When the butter melts, add the chicken, skin side down. Cook until the skin is crisp and golden, about 5 minutes. With tongs, turn the pieces and cook for 3 minutes longer. Remove from the pan and set aside.

4. Pour off all but ¼ cup fat from the pan. Add the carrots and onion. Cook, stirring often, for 10 minutes. Add the ciders and scrape up any sediment in the pan. Add the potatoes, increase the heat to high and boil vigorously for 5 minutes.

5. Add the stock, bacon, thyme, garlic, and chicken pieces, skin side up. Return to a boil. Transfer the pan to the lower third of the oven. Braise for 45 minutes, or until the chicken comes away easily from the bone. Serve hot.

INGREDIENTS

- 1 whole chicken, cut into 10 pieces
  Kosher salt and freshly ground black pepper
  Flour, for dredging
- ½ pound bacon, diced
- 2 tablespoons butter
- 3 carrots, cut into 1-inch lengths
- 1 medium onion, chopped
- 1 cup Norumbega Cidery Norumbega Classic Hard Cider
- ½ cup apple cider
- 3 medium potatoes, quartered
- 1 cup chicken stock
- 1 tablespoon chopped fresh thyme, plus sprigs for garnish
- 4 garlic cloves

THE PAIRING

Also based in rural Maine, Norumbega Cidery focuses on the tradition of American cider, starting with an orchard planted with cider apples and heritage varieties. The Norumbega Classic Hard Cider also adds some dessert apples into the mix, fermenting into a cider that is crisp, dry, and full of Maine apple flavor.

Woodchuck Cider Pulled Pork

Recipe by Woodchuck Hard Cider, Middlebury, VT

Woodchuck Hard Cider has been cooking with cider for decades, testing and trialing fermented batches with different flavor combinations in the kitchen. In this simple recipe for pulled pork, Woodchuck's Gumption cider proves itself a worthy braising liquid and ally in pairing with the pork. Top with the cider-soaked caramelized onions, shredded cheddar cheese, or a favorite coleslaw.

DIRECTIONS

Makes 6–8 servings

1. Heat 2 tablespoons of the oil in a large saucepan over medium-high heat. Sear the pork shoulder in the pan until browned on all sides, 8 to 10 minutes. Meanwhile, heat the butter and the remaining 2 tablespoons oil in a skillet over medium heat, and sauté the garlic and onions until caramelized, 5 to 7 minutes.

2. Once the pork is browned, place it in a slow cooker. Add the garlic and onions and pour 1½ bottles of the cider into the cooker. Turn the heat to low and add ½ cup of the barbecue sauce. Cook for about 8 hours or overnight.

3. Once the pork is cooked, remove from the cooker and shred the meat. Place the shredded meat back in the cooker, add the remaining 1½ cups barbecue sauce, and cook on low for 1 hour, then keep warm on the serve setting. Serve on a bun topped with the barbecue-onion sauce, cheddar cheese, or coleslaw, if using.

INGREDIENTS

- 4 tablespoons olive oil
- 1 4–6-pound pork shoulder
- 1 tablespoon butter
- 3 garlic cloves, sliced
- 2 large yellow onions, sliced
- 2 (12-ounce) bottles Woodchuck Hard Cider Gumption cider
- 2 cups barbecue sauce (your favorite brand)
- 6–8 buns
- 1 cup shredded cheddar cheese
- Coleslaw (optional)

THE PAIRING

Decades after the cidery launched, Woodchuck's ciders are seen on an international scale, including the Gumption, an honorary bottling for P. T. Barnum, the cider-loving showman behind Barnum & Bailey circus. Full of spunk and ingenuity like Barnum himself, the Gumption cider comprises dessert and cider apples and comes together in bittersweet and juicy apple flavors, an ideal yin and yang balance for this braise.

Lamb Shanks
Braised in ACE Cider BlackJack 21

Recipe by Executive Chef Sean Paxton, Home Brew Chef, Sonoma, CA

Finding his culinary muse in booze, Sean Paxton has been combining the flavors of beer and food for years under Home Brew Chef, an online community where chefs and brewers — amateur and pro — can collaborate and "rethink how beer is used in the kitchen and at the table." He brings the same sentiment to this recipe using ACE Cider BlackJack 21, a single-varietal Gravenstein cider from his home of Sonoma. "This recipe builds on flavors to enhance the Gravenstein apple essence in the hard cider with fennel, leeks, and shallots," Paxton explains. "The lamb shanks bathe in these flavors as they slowly braise, making the meat incredibly tender, sticky moist, and very rich and decadent, without being too heavy."

◈▷ DIRECTIONS

Makes 4 servings

1. Preheat the oven to 300°F (150°C). Wash the lamb shanks, removing any loose fat and blood, then blot dry with paper towels. In a shallow bowl, combine the flour with salt and pepper to taste; add the lamb and toss to coat evenly.

2. Warm 3 tablespoons of the oil in a 12-quart Dutch oven over medium-heat. Add 2 shanks to the pan, browning on all sides, about 4 minutes per side. Remove the shanks to a bowl and brown the remaining 2 shanks.

3. While the shanks are cooking, combine the leeks, fennel, shallots, bay leaves, and thyme in a bowl. Once the shanks are browned, check the oil in the pan. If it has burnt, remove it and add fresh oil. If the oil is fine, add the leek mixture and sauté until the vegetables just start to caramelize, 6 to 8 minutes. Season with 1 tablespoon kosher salt as the veggies cook down and intensify in flavor. Nestle the reserved lamb shanks

◈▷ INGREDIENTS

- 4 lamb shanks
- 2 tablespoons all-purpose flour
  Salt and freshly cracked black pepper
- 5 tablespoons olive oil, or more as needed
- 2 large leeks, cut in half, washed, and sliced, white and light green parts only
- 2 fennel bulbs, chopped, reserving the fronds for garnish
- 2 shallots, chopped
- 2 bay leaves
- 1 tablespoon fresh thyme
- 1 tablespoon kosher salt
- 1 (750 ml) bottle ACE Cider BlackJack 21 cider
- 2 Gravenstein apples, peeled and cut into ¼-inch cubes
- 1–2 teaspoons fennel pollen

into the vegetables, then pour in the cider. Gently shake each shank with tongs to let everything settle, checking the liquid level. The shanks should be three-quarters submerged in liquid; add water to achieve this amount, if needed. Cover and place the Dutch oven in the center of the oven. Braise for 3 hours.

4. When there are 15 minutes remaining for the shanks, sauté the apples with the remaining 2 tablespoons oil in a skillet over medium heat. Brown on all sides, lightly seasoning with salt.

5. Remove the lamb from the oven and check for doneness: The meat should still be attached to the bone, but tender enough to remove without a knife. Plate each shank, adding more seasoning, if needed, and spoon the braising liquid over it. Add caramelized apple cubes, then garnish with the reserved fennel fronds and a healthy pinch of fennel pollen. Serve immediately.

THE PAIRING

British expat Jeffrey House launched the California Cider Company and ACE Cider in 1993, making it one of the West Coast's first commercial cideries. A vintage cider to commemorate 21 (and aging) years of cidermaking, the BlackJack 21 ferments Gravenstein apples, barrel ages in Chardonnay barrels, and bottles as a cider that bubbles like sparkling wine. Taking the full bottle to the braise allows the lamb to really soak in the flavors of this full-bodied sipper.

Cider-Brine Quail
with Chilled Black-Eyed Pea Salad

Recipe by Chef Daniel Doyle, Poogan's Smokehouse, Charleston, SC

Poogan's Smokehouse is more than a barbecue joint: the varied plates strive to preserve the sanctity of traditional Southern cuisine. With inventive offerings and quintessential flavors found in the South, managing chef and partner Daniel Doyle uses the vibrant acid and sugar of Windy Hill Orchard's Gala Peach cider to brine this whole quail, then tosses it on the grill to get the char one would expect from a smokehouse recipe.

 DIRECTIONS

Makes 4 servings

1. Prepare the brine: In a large pot, combine the cider, water, salt, garlic, and thyme. Bring it to simmer, then chill. Submerge the quail in brine for 6 to 8 hours.

2. Prepare the black-eyed pea salad: Preheat the oven to 350°F (180°C). On a baking sheet, toss the tomatoes in the oil, season with salt and pepper, and roast for 7 minutes, until the skin starts to blister. Remove and let cool. In a large pot, bring the water and ham hock to a boil; add the peas and cook until tender. Strain the peas and lay out on a sheet tray to cool. Add the roasted tomatoes, chives, and celery to the tray and toss to combine, then chill.

3. Rinse the quail and pat dry. Preheat a grill, or warm a cast-iron pan over medium-high heat. Grill or cook the quail, turning occasionally to cook all sides evenly, until it reaches an internal temperature of 150°F (65°C), 4 to 6 minutes.

4. Prepare the mustard sauce: Combine the mustard, vinegar, sugar, honey, onion powder, garlic powder, cayenne, and salt in a bowl; taste for seasoning and adjust as needed.

INGREDIENTS

BRINE

- 2 cups Windy Hill Orchard and Cider Mill Gala Peach cider
- 1 cup water
- 3 tablespoons kosher salt
- 4 garlic cloves, smashed
- 3 sprigs thyme
- 1 whole (3–4 ounces) quail

BLACK-EYED PEA SALAD

- 10 cherry tomatoes
- 1–2 tablespoons olive oil
  Salt and freshly ground black pepper
- 1 gallon water
- 1 smoked ham hock
- 1 pound black-eyed peas
- 1 bunch chives, finely chopped
- 2 celery stalks, diced

MUSTARD BARBECUE SAUCE

- 1 cup yellow mustard
- ½ cup cider vinegar
- 3 tablespoons brown sugar
- ¼ cup honey
- ½ teaspoon onion powder
- ½ teaspoon garlic powder
  Dash of cayenne pepper
  Pinch of kosher salt

5. Mix half of the mustard sauce with the chilled salad, then brush ½ cup of the remaining sauce over the resting quail. Let the glazed quail sit for 5 to 10 minutes before carving, plate, and serve with the salad.

⬦ THE PAIRING

A tenured U-pick orchard and York, South Carolina, resident, Windy Hill Orchard and Cider Mill has been producing cider since 1996. The cider mill recently ramped up production to supply an increasing demand for such ciders as the Gala Peach, where apple and peach juices are fermented to dryness and fresh peach juice is added into the finished cider to replenish the tender peach flavor. The sweetness and acid are model attributes for the brine and rival the juiciness of the quail once it's off the grill.

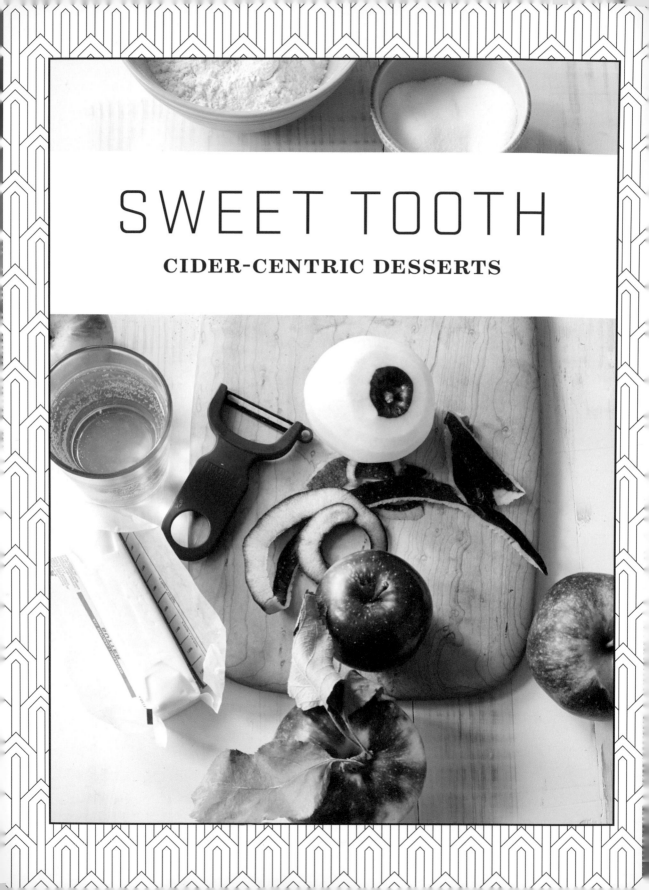

SWEET TOOTH

CIDER-CENTRIC DESSERTS

Rhubarb and Pistachio Tart

Recipe by Pastry Chef Dionne Himmelfarb, Ethan Stowell Restaurants, Seattle, WA

Dionne Himmelfarb's first restaurant job was at the Nordstrom Grill, within the flagship store of the Seattle retail luminary. Since then, she has moved up the ranks, learning the art of pastry from Northwest restaurant big shots Canlis and Poppy. She holds the reins for dessert production at Ethan Stowell Restaurants, a chef-run group of a dozen restaurants in the Emerald City. This rhubarb and pistachio tart embodies what Himmelfarb does best — mesh sweet with savory in a cider-friendly form.

DIRECTIONS

Makes 1 (11-inch) tart

1. Prepare the rhubarb filling: Combine the water, sugar, lemon zest, and cinnamon stick in a medium saucepan and bring to a boil over medium-high heat. Place the rhubarb in a medium bowl and carefully pour the hot syrup over it; cover the bowl. Let sit until the rhubarb is warm and slightly softened, 10 to 15 minutes. Drain the rhubarb and set aside to cool.

2. Prepare the pistachio filling: Combine the nuts, sugar, flour, eggs, butter, and vanilla in a food processor and process until it becomes a smooth nut batter. Set aside.

3. Preheat the oven to 300°F (150°C). Lightly butter an 11-inch tart pan with a removable bottom.

4. Prepare the tart shell: In a stand mixer fitted with a paddle attachment, beat together the butter and sugar on low speed until smooth, about 1 minute. Add the egg yolks and mix for another minute. Add the flour and salt, and mix just until a dough has formed. Press the dough evenly across the bottom and sides of the prepared pan. Spread the pistachio filling evenly on the tart shell. Bake for about 30 minutes, until

Recipe continues on page 260

INGREDIENTS

RHUBARB FILLING

- 4 cups water
- 4 cups sugar
- Zest of 1 lemon
- 1 cinnamon stick
- 3 cups chopped rhubarb

PISTACHIO FILLING

- 1½ cups pistachios (reserve ¼ cup, coarsely chopped, for garnish)
- 1 cup sugar
- 4 tablespoons all-purpose flour
- 2 large eggs
- ½ cup (1 stick) unsalted butter, at room temperature
- 2 teaspoons vanilla

TART SHELL

- 10 tablespoons unsalted butter
- ⅓ cup sugar
- 2 egg yolks
- 2¼ cups all-purpose flour
- ¼ teaspoon salt

the crust is golden and the filling is set. Let cool for about 10 minutes, then gently remove the tart shell ring and allow the pastry to cool completely.

5. To assemble, arrange the rhubarb filling on top of the pistachio filling, then garnish with the reserved chopped pistachios.

THE PAIRING

Westcott Bay Cider is Washington State's second-oldest cidery and also stakes claim as one of the region's oldest orchards. In a collaboration with its on-site sister distillery, San Juan Island Distillery, the cidery blends fresh apple juice with apple eau-de-vie and ages it in white oak barrels, resulting in the Westcott Bay Pommeau. The cider-based aperitif is the ideal partner for pairing with this tart, bestowing sweetness and acidity with that of the rhubarb and pistachio filling.

Hard Apple Cider Crisp

Recipe by Margo Greenman, Tacoma, WA

Food writer, amateur chef, and cider enthusiast Margo Greenman revels in experimenting in the kitchen and sharing her culinary prowess with others. From classic dishes turned up with modern flair to unique and one-of-a-kind creations, she takes an inventive and thoughtful approach to everything she makes, bakes, and cooks. Updating a family recipe, Greenman uses the basic principles of crisping apples and spikes them with local cider.

DIRECTIONS

Makes 1 (8- by 8-inch) crisp

1. Preheat the oven to 350°F (180°C).

2. Prepare the filling: In a mixing bowl, combine the apples, sugar, cinnamon, nutmeg, and cider. Mix well, then spread across an 8- by 8-inch baking pan.

3. Prepare the topping: In another bowl, combine the flour, granulated sugar, brown sugar, salt, and oatmeal. Using a pastry cutter or a fork, cut the butter into the flour mixture. Once well incorporated, spread the topping across the apple mixture.

4. Bake for about 45 minutes, until the top is slightly golden and the filling is bubbling. Let the crisp cool slightly before serving, and feel free to pair with a generous scoop of French vanilla ice cream.

INGREDIENTS

FILLING

- 4 apples peeled, cored, and cut into wedges ¼–½-inch wide
- ⅓ cup granulated sugar
- 2 teaspoons ground cinnamon
- 1 teaspoon ground nutmeg
- ½ cup Orondo Cider Works Dry cider

TOPPING

- 1 cup all-purpose flour
- ¼ cup granulated sugar
- ½ cup firmly packed light brown sugar
- ½ teaspoon salt
- ½ cup oatmeal
- ½ cup (1 stick) cold butter, cut into ½-inch cubes Vanilla ice cream, for serving

THE PAIRING

Orondo Cider Works has been pressing sweet cider on its Orondo, Washington, mill since 1980, recently adding hard cider to its production. The Orondo Cider Works Dry provides powerful, fragrant apple aromas and a juicy, crisp palate that matches the sweetness of the baking spices but follows the lead of the apple flavors in the crisp.

Maple Bread Pudding
with Cider-Soaked Apples

Recipe by Angry Orchard Hard Cider, Cincinnati, OH

The test kitchens at Angry Orchard are hot and cookin', playing with cider in everything from savory and grilled to sweet and baked. Marrying the flavors of maple and apple two ways, this divine, syrup-soaked bread pudding is a staff favorite and baking go-to for everything from the holidays to a lazy Sunday morning. This recipe requires an overnight bath for Fuji apples soaking in the cider.

DIRECTIONS

Makes 4 servings

1. Whisk together the eggs, cream, sugar, 1 cup of the maple syrup, vanilla, and maple flavoring in a large bowl. Add the soaked apples and brioche; stir to coat. Butter a 9- by 13- by 2-inch baking dish. Transfer the bread mixture to prepared dish and let stand at room temperature for 1½ hours.

2. Preheat the oven to 375°F (190°C). Bake for 40 minutes, until golden brown, slightly risen in the center, and a toothpick inserted in the center comes out clean.

4. Remove from the oven and, using a small knife or skewer, poke holes all over the pudding. Pour the remaining 1 cup maple syrup over the pudding and allow it to cool until just warm. Top with vanilla ice cream and serve.

INGREDIENTS

- 8 large eggs
- 1 quart heavy cream
- 1 cup granulated sugar
- 2 cups dark maple syrup, divided
- 2 teaspoons vanilla extract
- 1 teaspoon pure maple flavoring
- 1 cup Fuji apples, peeled, diced, and soaked overnight in Angry Orchard Crisp Apple cider
- 1 pound brioche, cut into 2-inch cubes
  Vanilla ice cream, for serving

THE PAIRING

The first cider for many United States citizens, Angry Orchard is committed to sourcing and growing quality ingredients for its ciders, starting with the Crisp Apple. This original sipper merges the juices of American dessert apples with those of French bittersweets of scintillating monikers such as Binet Rouge and Amere de Berthecourt. The blend is ripe, robust, and semisweet, a gateway drug for many and a plentiful provider of ample acid and juicy fruit for this bread pudding.

Perry-Poached Pear Frangipane

Recipe by Sean Kelly, WildCraft Cider Works, Eugene, OR

"Frangipane has been a longtime favorite for me," says Sean Kelly, owner, cidermaker, and resident chef at WildCraft Cider Works. "I have not always been fond of desserts, but this one has always struck a chord. . . . It helped push me through the long days and nights it took to get the cidery off the ground." A fan of frangipane and coffee (or cider), Kelly has crafted this recipe as an ode to the classic French dessert.

CRUST
- 6 tablespoons European-style unsalted butter
- 1 tablespoon vegetable oil
- 3 tablespoons evaporated cane sugar
- ⅛ teaspoon salt
- 1 cup all-purpose flour

PEARS
- 2 small, firm pears, like Bosc, sliced lengthwise into halves
- 1 (750 ml) bottle WildCraft Pioneer perry
- ½ cup evaporated cane sugar

FILLING
- ½ cup evaporated cane sugar
- ½ cup (1 stick) unsalted butter
- 1 egg
- ½ teaspoon almond extract
- 1½ ounces brandy
- 1 cup almond flour or finely ground blanched almonds
- 1 tablespoon all-purpose flour
  Mascarpone or whipped cream, for garnish

 DIRECTIONS

Makes 1 (11-inch) tart

1. Preheat the oven to 410°F (210°C). Lightly butter an 11-inch tart pan with a removable bottom.

2. Prepare the crust: Combine the butter, oil, sugar, and salt in a glass baking pan and bake for about 4 minutes, keeping a close eye until the mixture browns around the edges. Remove from the oven and briskly whisk in the flour. The mixture will resemble large cookie crumbs, not a dough ball or clump. Transfer the dough to the tart pan, reserving about 2 teaspoons to fill in the cracks that form during baking.

3. Once the dough is cool enough to handle, form the rim of the crust by pressing the crust together firmly and working around the pan and then through the center. This is a very crumbly mixture, so pressing together to form a continuous crust is important. With a fork, make a few punctures in the bottom to allow heat to escape. Bake for 10 to 15 minutes, until browned, checking every 2 minutes to see if cracks form. Fill in any cracks with the reserved dough. Once browned, cool the crust in the freezer.

Recipe continues on page 268

4. Meanwhile, prepare the pears: Gently core the pears with a spoon. Combine the perry and sugar in a stainless steel pot and bring to a simmer. Place the pears in the liquid, covering them with parchment paper. Poach for roughly 10 minutes, checking the firmness with a fork. Poaching time will vary depending on the ripeness and variety of pear.

5. Remove the pears with ¼ cup of the poaching liquid and cool in the fridge before handling. Continue to reduce the remaining poaching liquid, covered with parchment paper, at a low simmer for about 15 minutes or until the final liquid is equal to roughly ⅔ cup. Transfer into a dispensing container for easy use (like a squeeze bottle) to cool completely.

6. Prepare the filling: In the bowl of an electric mixer, combine the sugar, butter, egg, almond extract, and brandy until smooth. Add the almond flour and all-purpose flour, and blend to a paste.

7. To assemble, remove the cooled pears and crust from the fridge. Using a spoon or spatula, spread the filling evenly across the crust. Thinly slice the pears into cross-sections and arrange the slices splayed from the center out, with the skin facing up. The filling will rise up around them as it bakes.

8. Bake at 410°F (210°C) for 15 to 20 minutes, until the surface is completely browned. Remove from the oven, brush with the reserved cooled poaching liquid, and continue baking for another 5 to 10 minutes. When done, cool slightly and let set in the fridge for 3 to 4 hours before removing from the tart pan. The crust is very flaky and can crumble if not handled gently. Slice with a sharp knife, top with a dollop of mascarpone, and serve.

THE PAIRING

WildCraft's Pioneer Perry is a blend of pears, cultivated and wild, hand-picked from the original homesteaded farms in Lane County, Oregon. The feral fruit contributes unique flavors and tannin while maintaining delicacy and earthy pear flavors. Hazelnut and shortbread flavors meld with the tastes of the pastry, making it a natural fit to cook the pears in and to match the crust.

READING LIST

Apples of North America: Exceptional Varieties for Gardeners, Growers, and Cooks by Tom Burford (Timber Press, 2013)

Apples of Uncommon Character: Heirlooms, Modern Classics, and Little-Known Wonders by Rowen Jacobsen (Bloomsbury USA, 2014)

Apples to Cider: How to Make Cider at Home by April White and Stephen M. Wood (Quarry Books, 2015)

The Botany of Desire: A Plant's-Eye View of the World by Michael Pollan (Random House, 2001)

Cider, Hard and Sweet: History, Tradition, and Making Your Own by Ben Watson (Countryman Press, 2013)

Ciderland by James Crowden (Birlinn Ltd., 2008)

Cider Made Simple: All About Your Favorite New Drink by Jeff Alworth (Chronicle Books, 2015)

Craft Cider Making by Andrew Lea (Good Life Press, 2011)

The Holistic Orchard: Tree Fruits and Berries the Biological Way by Michael Phillips (Chelsea Green Publishing, 2012)

The New Book of Apples: The Definitive Guide to over 2,000 Varieties by Joan Morgan and Alison Richards (Ebury Press, 2003)

The New Cider Makers Handbook by Claude Jolicoeur (Chelsea Green Publishing, 2013)

Old Southern Apples: A Comprehensive History and Description of Varieties for Collectors, Growers, and Fruit Enthusiasts by Creighton Lee Calhoun Jr. (Chelsea Green Publishing, 2011)

World's Best Cider: Taste, Tradition, and Terroir by Pete Brown and Bill Bradshaw (Sterling Epicure, 2013)

ACKNOWLEDGMENTS

From Erin James: Every morning when I wake up, I am thankful for my husband, Nick James. I couldn't have completed this book without his enduring advocacy, patience, and calming presence. I would also like to thank my friends and family for their encouragement and faith during this process, even when I became a recluse hunched over a laptop. A standing ovation to my trusted and supportive business partners Kristin Ackerman Bacon and Melissa Miller, along with the rest of the stellar team at CIDERCRAFT magazine. I love you!

From CIDERCRAFT: We would like to thank the following people for their support, assistance, and overall greatness in making this book possible: Clelia Gore, Jill Lightner, Julia Wayne, Amy Johnson Photography, Brittany Carvalho/Blue Rose Photography, and Lindsay Borden Photography.

The cultured, brilliant, and generous cidermakers who took the time to sit down with us and chat cider (in order of appearance): Stephen Wood, Autumn Stoscheck, Kevin Zielinski, Bruce Nissen, Joel VandenBrink, Ryan Burk, Chuck Shelton, Field Maloney, François Pouliot, Kristen Needham, James Kohn, Tim Edmond, Michel Jodoin, Tim Larsen, Mike Beck, Trudy Davis, Chris Haworth, Scott Donovan, Crystie Kisler, Sean Kelly, Nat West, Dave Takush, Charles McGonegal, Jim Gerlach, and Marcus Robert.

A special thank-you also to the participating cideries, bars, and restaurants that donated their delicious, cider-savvy recipes and time to be featured in this book.

And last, but certainly not the least, we have immense gratitude for Carleen Madigan, Alethea Morrison, Mars Vilaubi, and the rest of the talented team at Storey Publishing for bringing this dream of ours to life.

INDEX

PHOTO CREDITS

TREAT YOUR TASTE BUDS
WITH MORE BOOKS FROM STOREY

by Olwen Woodier

Apple pie is just the beginning. Discover the versatility of this iconic fruit with 125 delicious recipes for any meal, including apple frittata, pork chops with apple cream sauce, apple pizza, apple butter, and more.

by Randy Mosher

Uncap the secrets in every bottle of the world's greatest drink. Learn what makes each brew unique and how to identify the scents, colors, flavors, and mouthfeel of nearly 90 beer styles from around the globe.

by Lew Bryson

This comprehensive guide to tasting and enjoying the world's whiskeys includes Scotch and bourbon as well as Tennessee, Irish, Japanese, and Canadian whiskey. Settle in for the journey with a cocktail made from one of the book's recipes.

by Ellen Zachos

Create delicious mixed drinks using common flowers, berries, and roots. Get your party started with more than 50 recipes for garnishes, syrups, juices, and bitters, then incorporate your handcrafted components into 45 cocktails such as Stinger in the Rye, Don't Sass Me, and Tree-tini.